Optoelectronics

Edited by Mike Haidar Shahine

Published in London, United Kingdom

IntechOpen

Supporting open minds since 2005

Optoelectronics
http://dx.doi.org/10.5772/intechopen.87841
Edited by Mike Haidar Shahine

Contributors
Mike Haidar Shahine, Hemali Rathnayake, Sheeba Dawood, Putha Kishore, Manchineellu Padmavathi, Dinakar Dantala, Andrey G. Georgievich Paulish, Peter S. Zagubisalo, Sergey M. Churilov, Vladimir N. Barakov, Mikhail A. Pavlov, Alexander V. Poyarkov, Nikhil Deep Gupta, Feng Li, Senthilkumar Chandran, Srinivasan Manikam, Thamer A. Tabbakh, Zahrah Abdullah Almusaylim, Sundos M. Alabdulkarim, Noha Alotabi, N.Z Jhanjhi, Nawaf Bin Darwish

First published in London, United Kingdom, 2021 by IntechOpen
IntechOpen is the global imprint of INTECHOPEN LIMITED, registered in England and Wales, registration number: 11086078, 5 Princes Gate Court, London, SW7 2QJ, United Kingdom
Printed in Croatia

British Library Cataloguing-in-Publication Data
A catalogue record for this book is available from the British Library

Additional hard and PDF copies can be obtained from orders@intechopen.com

Optoelectronics
Edited by Mike Haidar Shahine
p. cm.
Print ISBN 978-1-83962-400-1
Online ISBN 978-1-83962-401-8
eBook (PDF) ISBN 978-1-83962-402-5

For EU product safety concerns:
IN TECH d.o.o., Prolaz Marije Krucifikse Kozulić 3, 51000 Rijeka, Croatia, info@intechopen.com or visit our website at intechopen.com.

Meet the editor

Dr. M. H. Shahine is affiliated with General Dynamics Mission Systems (GDMS). Prior to GDMS, he had leading roles at Acacia Communications, WiConnect Global Inc., and Ciena Corporation. Dr. Shahine served as a member of the board of directors for the Optoelectronic Industry Development Association (OIDA) in Washington, DC. He represented the Ciena Corporation from 2007 until 2012. He is also an adjunct professor at the University of Maryland system and Harrisburg University. He received an MSc. degree from the Johns Hopkins University and a Ph.D. from the University of Maryland, both in electrical engineering. He is a senior member of the IEEE. He holds nine issued and several pending patents. He is the author of a book chapter on Semiconductor Laser Diode.

Contents

Preface

Optoelectronics is one of the fast-emerging, technological fields that deals with applying electronic devices to the sourcing, detection, and control of light. This technology is used for numerous purposes such as telecommunications, monitoring and sensing, long-wavelength LiDAR, microwave photonic links, medical equipment, fast computing, and general science.

This book describes the state of the art in this rapidly evolving field. It provides a broad spectrum of technical information and research ideas in both applications and material research. It deals with advanced applications in its first section, and with materials and structures in its last section. The individual chapters were written by leading scientists in their field, with the hope that this book will help students and active researchers who are interested in the physics and applications of optoelectronics.

The first section of the book deals with applications that are currently of wide interest, starting with chapter one that covers neuromorphic photonics and highlights the use of semiconductor lasers as neurons. Neuromorphic photonics is a neuroscientific concept of introducing optoelectronic feedback to control semiconductor lasers that mimic the neuron-firing process in the brain and encode information from those pulses for efficient signal processing. Chapter two covers biosensing applications using the optical surface plasmon resonance effect. It highlights current state-of-the-art sensing techniques for various materials and the fabrication techniques for the sensing devices. Chapter three covers piezo-optical transducers for highly sensitive strain measurements and the process for creating a strain gauge that is more highly sensitive than other techniques for mechanical stresses. Chapter four covers noncontact vibration sensors based on the use of plastic optical fiber. This novel sensing technique eliminates some of the significant effects such as power fluctuation in light sources and bending loss.

Section two of this book covers research on materials and structures. Chapter five covers highly efficient solar cells using periodic nanostructures based on light management for efficient solar energy harvesting. Chapter six provides a detailed discussion on hybrid perovskite crystals with different dimensional forms of bulk and thin crystals. It also lists the potential benefits for use in future optoelectronic devices. Chapter seven examines and analyzes in detail the optical and electronic properties of single crystals. This work provides a valuable guide for developing single crystal–based material. Lastly, chapter eight covers a detailed analysis of polymer, MOF material for use in next-gen optoelectronic devices. It highlights particularly interesting properties, such as tunable bandgap and varied magnetic properties, among other things.

Finally, I want to thank the authors for their valuable contribution to this book. Their work is greatly appreciated, especially the extra efforts made during the global Covid-19 pandemic to get this book published. I also would like to thank Ms. Marijana Francetic for her help with the IntechOpen publication process.

Dr. Mike Haidar Shahine
General Dynamics Mission Systems,
Maryland, USA

Section 1

Advanced Applications in Optoelectronics

Chapter 1

Neuromorphic Photonics

Mike Haidar Shahine

Abstract

Neuromorphic photonic applies concepts extracted from neuroscience to develop photonic devices behaving like neural systems and achieve brain-like information processing capacity and efficiency. This new field combines the advantages of photonics and neuromorphic architectures to build systems with high efficiency, high interconnectivity and paves the way to ultrafast, power efficient and low cost and complex signal processing. We explore the use of semiconductor lasers with optoelectronic feedback operating in self-pulsating mode as photonic neuron that can deliver flexible control schemes with narrow optical pulses of less than 30 ps pulse width, with adjustable pulse intervals of −2 ps/mA to accommodate specific Pulse Position Modulation (PPM) coding of events to trigger photonic neuron firing as required. The analyses cover in addition to self-pulsation performance and controls, the phase noise and jitter characteristics of such solution.

Keywords: neuromorphic, neuron, optoelectronic feedback, photonic integration, self-pulsation, control theory, semiconductor laser, photonic tensor core

1. Introduction

Von Neumann digital computer architecture [1] that existed since the 1940s and still being the only viable architecture for computers cannot keep up with the exponential speed needed, to process data for machine learning and artificial intelligence applications as we move into a internet of things (IoT) dominated world. This architecture cannot keep up with Moore's law that predicted the count of transistors in a CPU to double every 2 years, while at the same time, the CPU clock rate reached a ceiling at 4 GHz due to prevalence of current leakage in nanometric nodes. Hence, the move to multicore architecture is running against the power requirements for simultaneously powering these cores. All this can be traced to the excess amount of energy consumption of digital switching and the bandwidth limitation of the metal interconnects. These listed bottlenecks are driving the efforts for a new computing architecture towards the use of neuromorphic photonics, especially with the fast track to maturity that photonic integration has taken with III-V material processing and recently with Silicon photonics. Photonic integration offers a rich library of various components with reduced latency, higher bandwidth, and energy efficiency. It also facilitates nonlinear optoelectronic devices along with photonic/electronic integration and compact packaging.

The chapter is organized as follows: Section 2 covers background information on Neurons, and the efficiency of information processing in the human brain compared with other available technologies, it also covers addresses photonic tensor cores, the basic architecture of photonic neuron, and how the information is coded. Section 3 introduces photonic neuron based semiconductor lasers with optoelectronic feedback, along with feedback control theory, and the equations covering self-pulsating mode of laser neuron system. Section 4, presents simulation results and noise analysis of the photonic neuron system, while Section 5, provides the concluding remarks of this work.

2. Technology survey

In this section, we survey competing technologies and their architectures while comparing them to the ultimate information processor in the human brain.

2.1 Neurons

The human brain contains around 100 billion neurons. It is the most complex system for information processing in existence, with an execution power of 10e18 (multiply-accumulate matrix) MAC/s with only 20 Watts power needed [2, 3]. Each of those neurons has an average of 10e4 inputs of tiny neurotransmitter junctions known as synapse. This translates into 10e15 of synapses connections, while the bandwidth of the signal processed in the brain is 1 KHz max. The computational efficiency of the brain is less than 1 AttoJoules/MAC while supercomputers today have an efficiency of 100 picoJoules/MAC, in other word, the magnitude of the brain efficiency is 8 orders better than supercomputers as can be seen in **Figure 1**. The neuron by definition is a single brain cell. It communicates with other neurons with electrical impulses via thousands of synapses. The neurotransmitters (chemical secretions) electrical input to the neuron comes from other neurons with different weights. The signal level needs to exceed a certain threshold value to cause the neuron to fire, sending a series of electrical spikes to other neurons. Below threshold input, the neuron output is very small and linear, while above threshold, the neuron output is large and nonlinear. This behavior is similar to how lasers operate; below threshold, the optical output is made of spontaneous emission, low output incoherent optical power, while above threshold, stimulated emission dominates and produce large coherent optical light output. A key requirement to have the laser behave as a neuron is to include an absorber section to limit the spontaneous emission output below threshold levels. These electrical activities in the brain can be monitored using ElectroEncephaloGraph (EEG), these signals can be

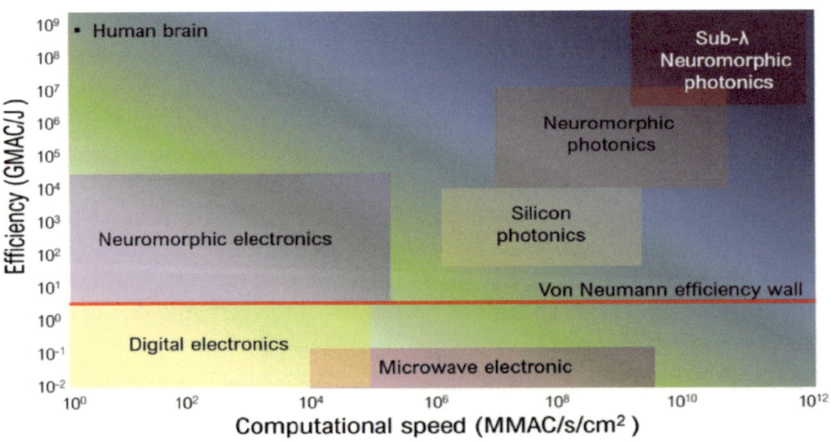

Figure 1.
Comparing computational speed with efficiency for various elements of available technologies [4].

strongly correlated with different cognitive tasks, but they appear as a chaotic signal plot when plotted against time from physics and mathematics perspectives.

2.2 Photonic tensor core

The key driver for neuromorphic photonic approach especially in Machine-Learning, is the move from electronic processing approach in hardware systems such as Google's Tensor Processing Unit (TPU) which relies on grinding through stacks of repeated matrix multiplications that require immense amount of power. Neuromorphic photonic approach facilitates a photonic solution that is a modular photonic tensor core (PTC) where all matrix calculations get processed in the optical domain. This PTC process provides three orders of magnitude more computing power than TPU with processing of the multiply-accumulate matrix (MAC) operations. TPU data also exhibit long run times especially when performing image processing. While PTC can use the wave nature of light to, directly perform summation using coherent addition of wave amplitudes, and multiplications can result from the interaction of optical waves with matter. At the same time, and similar to the behavior of biological neurons, where each neuron can both process and store data, PTC replicates these functions, which dramatically reduces latency [5].

2.3 Photonic neuron

The neuron consists of an input stage that is a linear combination (weighted addition) of the outputs of the neurons feeding it. The combined signals from the input stage are integrated to produce a nonlinear response as dictated by the activation function see **Figure 2**.

This neuron must perform three mathematical operations:

1. Vector (weighting) multiplication

2. Spatial summation

3. Nonlinear transformation

All inputs must be of the same nature as outputs.

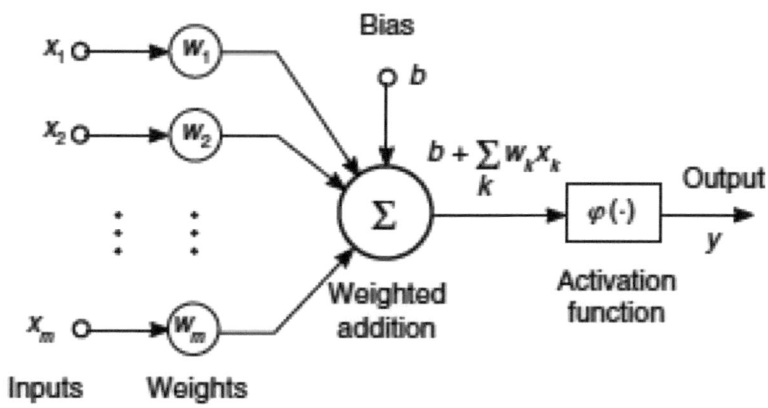

Figure 2.
Neuron signal flow model [6].

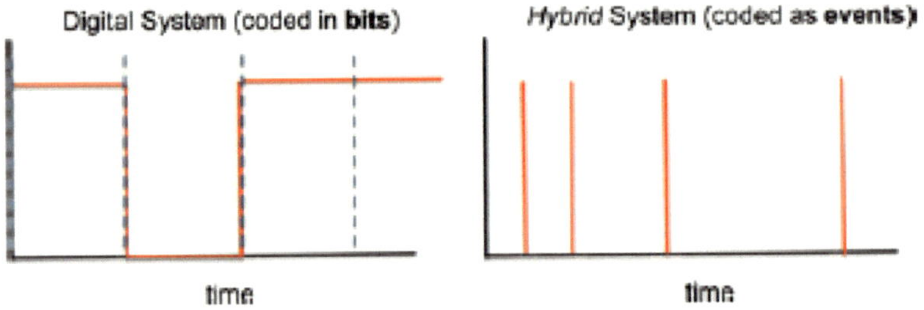

Figure 3.
Spiking neuron information encoding scheme of events rather than bits [6].

2.4 Information coding

In order to have an efficient signal modulation scheme, a hybrid combination of analog and digital signal representation need to be implemented that mimics the way the brain encodes information as events in time. The type of a well-known modulation scheme used in optical communication systems called Pulse Position Modulation (PPM) [6, 7].

This modulation format exploits the efficiency of analog signals and at the same time reduces the noise accumulation that distorts analog signals (see **Figure 3**).

3. Semiconductor lasers with optoelectronic feedback as photonic neuron

Semiconductor lasers are widely used in many applications for both digital and analog signal processing. For the past two decades, specifications of these lasers have addressed many specific applications by tailoring the laser design parameters to meet specific performance target. While the aim of this work is to use low cost lasers with generic specifications, modify and enhance their essential performance parameters to behave as a photonic neuron in addition to the applications addressed in [8] by using electronic feedback for triggering self-pulsating behavior necessary for spiking neuron model. The main driver behind this work is to facilitate photonic integration, improving laser modulation bandwidth and increasing laser relaxation oscillation frequency. The performance for all these applications is analyzed in both time and frequency domains. Mainly, by adjusting the feedback loop settings so it can operate outside its stable regime but just ahead of the chaos mode, so the laser can run in self-pulsation mode. This provides the use of the laser drive current as a single point of control for the pulsation rep-rate and.

3.1 Basic control theory

The theory behind this work is based on the classical control theory of negative feedback. Recent work by [9], has presented a rigorous, yet simple and intuitive, non-linear analysis method for predicting injection locking in LC oscillators.

A system with a negative feedback control loop is shown in **Figure 4**.

It consists of a forward-gain element with transfer function $A(s)$, with s is the Laplace operator and can be replaced by $(j\omega)$ feedback element with transfer function $B(s)$.

Where $A(s)$ represents laser transfer function, $B(s)$ represents the feedback loop transfer function, x is the injection Current and y is the Optical output power.

The closed loop transfer function of such system is:

$$T(s) = \frac{y}{x} = \frac{A(s)}{1 + A(s) * B(s)} \tag{1}$$

This system can be linearized by making the gain product $A(s) * B(s) \gg 1$.

With this condition, the transfer function becomes dependent solely on the feedback gain coefficient and response of the feedback loop, which becomes linear:

$$T(s) = \frac{1}{B(s)} \tag{2}$$

A feedback loop can oscillate if its open-loop gain exceeds unity and simultaneously its open-loop phase shift exceeds π. There are poles that are present in the closed loop configuration with at least one pole of an unstable loop lies in the right half of the s-plane in **Figure 5**.

Analysis of stability of this system can be done according to the Nyquist Criterion [10] by investigating pole location.

It is based on Nyquist plot where the open-loop transfer function is analyzed with a plot of real and imaginary parts. Where stability of the closed-loop system is determined If poles are present in the Left Half (LHP) of the s-Plane.

If poles are present in the Right Hand Plane (RHP), the closed-loop system becomes unstable. In brief, the Nyquist criterion is a method for the determination of the stability of feedback loop systems as a function of an adjustable gain and delay parameters in the feedback section. It simply determines if the system is stable for any specified value of the feedback transfer function $B(s)$.

For the self-pulsation mode, it is well known that an active circuit with feedback can produce self-sustained oscillations only if the criterion formulated by Barkhausen [10] is fulfilled. This criterion is defined as when the denominator of

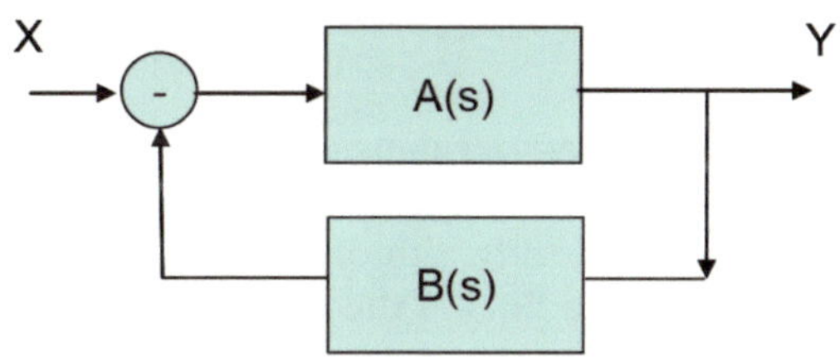

Figure 4.
Negative feedback loop system [8].

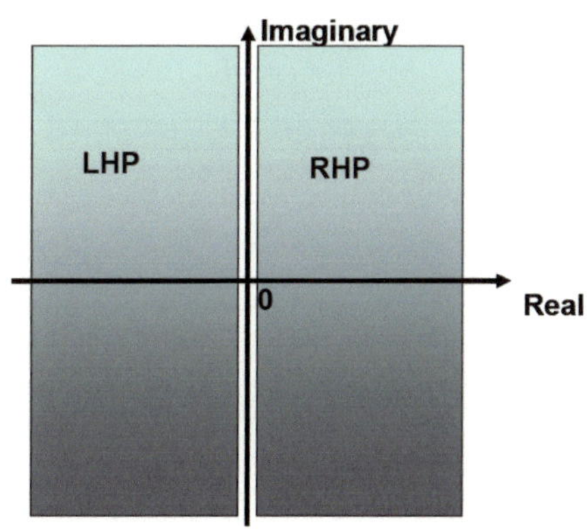

Figure 5.
S-plane plot.

the closed loop transfer function is zero. The poles in this self-pulsation mode need to be located outside RHP (Chaos mode) and LHP (Stable mode), only present up and down on the imaginary axis of the s-plane plot with a zero value on the real axis and where the phase of this transfer function:

$$\Delta T(\ j\omega) = 0 \Rightarrow \omega = \omega_0 \tag{3}$$

$$|T(j\omega_0| = 1 \tag{4}$$

Eqs. (3) and (4) are the phase and gain conditions, respectively. Based on Barkhausen criterion, the oscillation frequency is determined by the phase condition (3).

3.2 Self-pulsating system

There have been numerous publications on the effects of lasers with electronic feedback [11–13], covering mainly the various states from operating this system. The in this work is on increasing the feedback loop delay to achieve self-pulsation and chaos modes.

Starting with the DFB laser characteristics that are modeled using the well-known rate equations [8, 14] that have been modified to include the electronic feedback parameters.

$$\frac{dN(t)}{dt} = \frac{I(t)}{q * V_a} - g_0 \frac{[N(t) - N_0] * S(t)}{1 + \varepsilon * S(t)} - \frac{N(t)}{\tau_n} + \left[\frac{\omega_n}{2\pi} * (\rho * S[t - \tau])\right] \tag{5}$$

$$\frac{dS(t)}{dt} = \Gamma * g_0 \frac{[N(t) - N_0] * S(t)}{1 + \varepsilon * S(t)} - \frac{S(t)}{\tau_p} + \frac{\Gamma * \beta}{\tau_n} * N(t) \tag{6}$$

$$\frac{d\phi(t)}{dt} = \frac{1}{2}\alpha\left[\Gamma * g_0[N(t) - N_0] - \frac{1}{\tau_p}\right] \tag{7}$$

Where Eq. (5), represents the carrier density equation with the feedback terms, ρ represents the feedback loop gain, τ represents the feedback loop propagation delay and ω_n represents the 3 dB bandwidth of the amplifier circuit. Eq. (6), represents the photon density, and Eq. (7) the optical phase.

The laser parameters included in these equations are listed in **Table 1**.

Symbol	Value	Dimension	Description
I(t)	—	A	Laser current
S(t)	—	m^{-3}	Photon density
Γ	0.44	—	Optical confinement factor
g_0	$3x10^{-6}$	cm^{-3}/s	Gain slope
N(t)	—	m^{-3}	Carrier density
N_0	$1.2x10^{18}$	cm^{-3}	Carrier density at transparency
ε	$3.4x10^{-17}$	cm^3	Gain saturation parameter
τ_p	$1.0x10^{-12}$	s	Photon lifetime
β	$4.0x10^{-4}$	—	Spontaneous emission factor
τ_n	$3.0x10^{-9}$	s	Carrier lifetime
Va	$9.0x10^{-11}$	cm^3	Volume of the active region
Φ	—	—	Phase of the electric field from the laser

Symbol	Value	Dimension	Description
α	3.1	—	Line-width enhancement factor
P(t)	—	W	Optical power from laser
Q	$1.602x10^{-19}$	C	Electronic charge
η	0.1	—	Total quantum efficiency
h	$6.624x10^{-34}$	J.s	Plank's constant
ω_n	$75.4x10^{9}$	Rad/s	3 dB Bandwidth of amplifier Circuit
σ_g	$2x10^{-20}$	m^2	Gain cross section

Table 1.
Paramters used in this work [14].

The optical output power from the laser is represented in Eq. (8)

$$\mathbf{P(t)} = \frac{\mathbf{S(t)} * \mathbf{V_a} * \eta_0 * \mathbf{h} * \nu}{2 * \Gamma * \boldsymbol{\tau}_\mathbf{p}} \tag{8}$$

The system being analyzed which includes the laser and the feedback loop is shown in **Figure 6**.

This system consists of a wideband monitor diode located at the back facet of the laser cavity and electrical amplifier. This implementation using the wideband back-facet monitor [15], provides the means to control and manage the short propagation delay in the feedback loop, this is necessary layout in order to achieve the desired performance characteristics. It also provides a mechanically stable system. The key parameters calculated from the model equations are the relaxation oscillation frequency (ROF), and the damping factor. The system is configured to account for the delay, gain and bandwidth of the feedback loop and are expressed in the following forms:

$$f_r = \frac{1}{2\pi}\sqrt{K - \frac{1}{2}(\gamma_d)^2} \tag{9}$$

The laser transfer function H is of the form:

$$H(j\omega) = \frac{K}{[(j\omega) * (j\omega + \gamma_d)] + K} \tag{10}$$

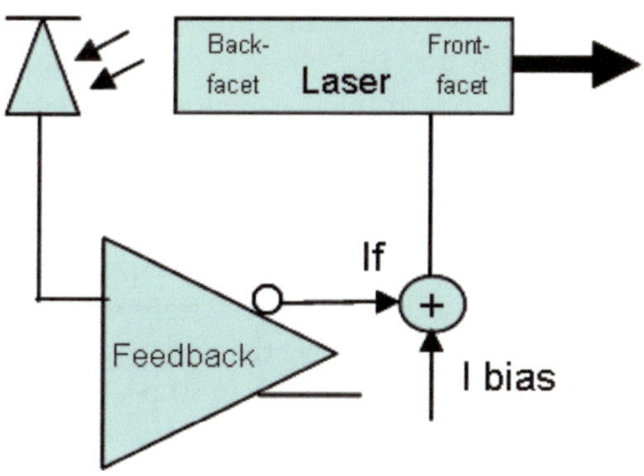

Figure 6.
Laser system with feedback.

Where K is:

$$K = \left[\frac{\Gamma * g_0}{q * V_a}(I_{Bias} - I_{th})\right]\left[1 - \frac{\Gamma}{q * V_a} * \varepsilon * \tau_p(I_{Bias} - I_{th})\right] \quad (11)$$

and γ_d, the damping factor, is of the form:

$$\gamma_d = \frac{1}{\tau_n} + \left[\frac{\Gamma * g_0}{q * V_a}\left(\tau_p + \frac{\varepsilon}{g_0}\right)((I_{Bias} - I_{th})\right]\left[1 - \frac{\Gamma}{q * V_a} * \varepsilon * \tau_p(I_{Bias} - I_{th})\right] \quad (12)$$

In the feedback loop, the amplifier transfer function A is of the form

$$A = \frac{-\rho}{1 + \left(\frac{j\omega}{\omega_n}\right)} \quad (13)$$

Where ρ is the feedback gain, and ω_n is the 3 dB bandwidth of the amplifier circuit.

For the delay transfer function B is of the form

$$B = e^{-j\omega\tau} \quad (14)$$

Where τ is the propagation time delay of the feedback loop system.

Based on the well-known control theory of systems with negative feedback [16], the complete transfer function on this complete laser system Y is of the form

$$Y(j\omega) = \frac{H}{1 + (H * A * B)} \quad (15)$$

Using the parameters listed in **Table 1**, the calculated laser threshold current is 9.4 mA. The slope efficiency is calculated at 0.04 mW/mA.

4. Simulation results and discussion

Setting up the system to operate in self-pulsating state with fixed FB loop gain of −0.05 and loop delay of 50 ps, the physical phenomenon of self-pulsation process is described as the sharpening and extraction of the first spike of the relaxation oscillation frequency (ROF) of the laser cavity. The feedback sharpens the falling edge of the first spike and suppresses the subsequent spikes.

Hence, lasers with stronger ROF generate shorter pulses. We show the system transfer function (Y(jω)) magnitude and phase plots in **Figure 7**. What we see is in the case where the feedback loop is applied an enhanced second peak in the magnitude transfer function plot of **Figure 7** which indicates the generation of sharp pulsation. The inverse of the frequency peak corresponds to the pulse interval in the time domain.

A capture of the time-domain picture of the self-pulsation mode, is shown in **Figure 8** where the set points are at 50 mA bias current with feedback delay of 50 ps and feedback gain of −0.05. This plot shows the output power of the system where the pulse interval is 147 ps and the pulse width is 50 ps.

Figure 9 shows the change of the pulse interval (Free Spectral Range) as a function of the bias current. The pulse interval can be fine-tuned over a range > 50 ps. The shortest pulse interval was achieved for these particular laser parameters from **Table 1**. When setting the delay at 30 ps and the gain at −0.05

with 50 mA bias current was 80 ps with pulse width of 30 ps. These limitations on the pulse width are governed mainly by the laser carrier lifetime in the laser structure. This is a crucial feature for photonic neurons, as the pulse interval can be adjusted by modulating the laser current, where asymmetric spacing is needed based on specific events that lead to neuron firing.

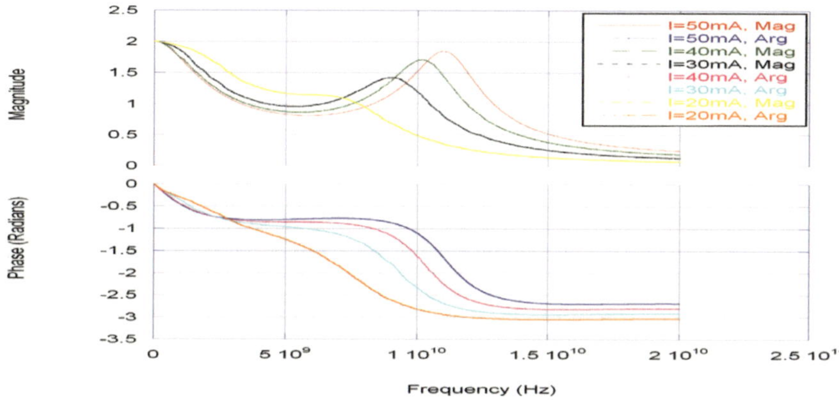

Figure 7.
Magnitude and phase plots of the laser transfer function with feedback loop in self-pulsation regime for various current values (FB loop gain = −0.05, FB loop delay = 50 ps).

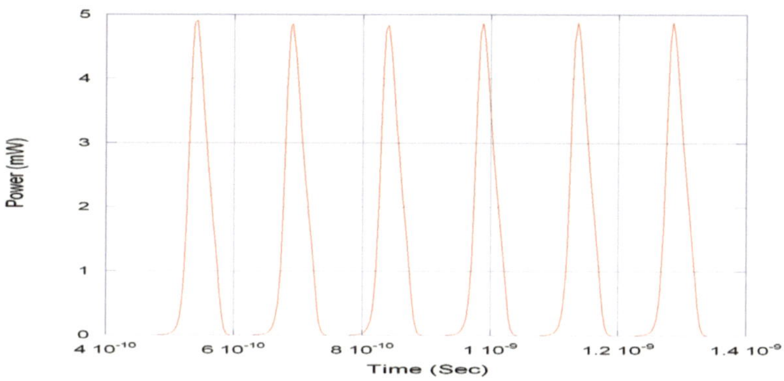

Figure 8.
Time domain plot for self-pulsation case (delay = 50 ps, gain = −0.05) for 50 mA bias current where the pulse interval is 147 ps.

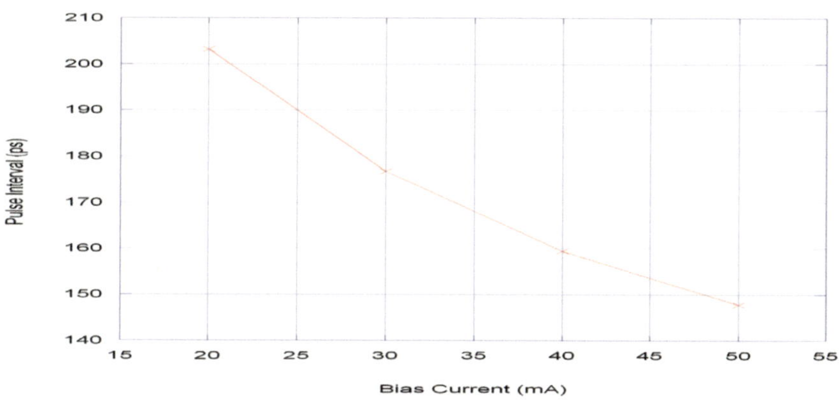

Figure 9.
Pulse interval adjustment as a function of bias current for 50 ps delay and gain of −0.05.

4.1 Pulsed-source noise analysis

In section (2) of this chapter, we analyzed the rate Eqs. (5)–(7) without the inclusion of the Langevin noise terms $F_N(t), F_S(t)$ *and* $F_\varphi(t)$ are the noise terms added respectively to the rate equations. These noise terms are Gaussian random processes with zero mean value under the Markovian assumption (memory-less system) [17]. The Markovian approximation of this correlation function is of the form:

$$\langle \mathbf{F_i(t)F_j(t')} \rangle = \mathbf{2D_{ij}\delta(t - t')} \tag{16}$$

Where $i,j = S,N,$ *or* ϕ.

D_{ij} is the diffusion coefficient with full derivation presented in [17].

The other type of noise effect analyzed is the system phase noise, which has dramatic effects on the performance of pulsed laser sources especially when it comes to timing jitter. The system phase noise $L(f)$ is produced from the effect of the laser linewidth $\delta\nu$ and the power spectral density $S_\varphi(f)$ of that linewidth.

$$S_\varphi(f) = \frac{1}{1 + \frac{2*f}{\delta\nu}} \tag{17}$$

The system phase noise $L(f)$ shown in **Figure 10** is related to the linewidth power spectral density as follow [18]:

$$L(f) = \frac{S_\varphi(f)}{2} \tag{18}$$

The integrated rms timing jitter σ_j which represents the upper bound of the timing jitter of the oscillator shown in **Figure 11** is calculated as follow [19].

$$\sigma_j = \frac{Pulse\ Interval}{2\pi}\sqrt{2 * \int_{fmin}^{fmax} L(f)df} \tag{19}$$

Where $f\,min$ and $f\,max$ are the boundary of the frequency range.

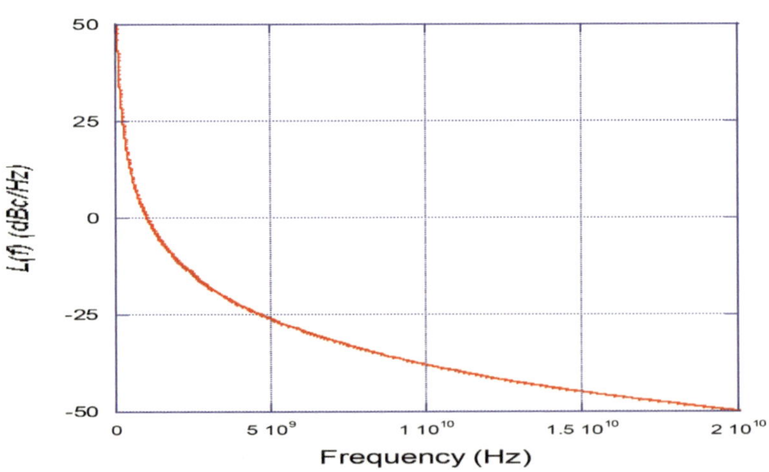

Figure 10.
Laser phase noise plot derived from the spectral density of the line-width.

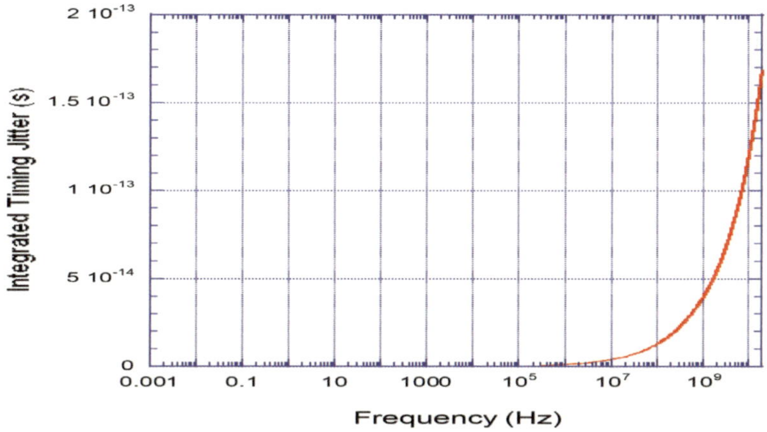

Figure 11.
rms timing jitter over the entire frequency range.

For a pulsed source with a pulse interval of 80 ps, the maximum tolerated rms jitter for sampling application is 120th the pulse interval according to [19]. The listed requirements of maximum tolerated rms jitter is 667 fs while our calculated jitter shown in **Figure 11** is around 15 fs.

We also analyzed how certain laser physical design parameters presented can enhance further the performance of this self-pulsating laser structure with feedback for neuromorphic application. Our analysis determined that increasing the laser cavity length can produce a narrower linewidth by increasing the photon lifetime which will enhance further the timing jitter performance, another approach is to use quantum dot based laser structures which can deliver close to zero or negative linewidth enhancement factor (α parameter).

5. Conclusion

Based on the modified rate equations for analyzing DFB laser system with electronic feedback, this work addresses the need for self-pulsating laser behaving as a photonic neuron, this work provides detailed requirements for feedback loop delay, bandwidth, and gain ranges required to operate the laser self-pulsation modes. These effects were simulated numerically and guidelines were generated for the list of recommended parameters necessary to realize such system. The time domain pulse interval which is crucial for neumorphic photonics was also analyzed using only the laser drive current for tuning the pulse interval of −2 ps/mA for the realization of variable spaced pulses necessary for this application with pulse spikes as narrow as 30 ps. We also provided analysis of phase noise and rms jitter. These results also show that a pulse train can be generated and controlled with only the laser bias current without the use of external clocking or signaling sources, while PPM signals can ride on top of the laser current modulation to code signals into the laser output, which now provides to degrees of adjustments, one for the pulse grouping (interval) and one for the information to be transmitted using PPM.

Author details

Mike Haidar Shahine
General Dynamics Mission Systems, Maryland, USA

*Address all correspondence to: dr.m.shahine@ieee.org

References

[1] Von Neumann J. IEEE Annals of the History of Computing. 1993;**15**:27

[2] Sporns O., Networks of the Brain, 1st Edition, ISBN 978–0–262-01469-4 (2011)

[3] Hasler J et al. Frontiers of. Neuroscience. 2013;**7**:118

[4] Goi E et al. Perspective on photonic memristive neumorphic computing, PhotoniX journal, springer open, Vol. 1. Issue. 2020:3

[5] Miscuglio M., et al., Applied Physics Reviews 7, 031404 (2020)

[6] Nahmias M et al. IEEE journal of selected topics in quantum electronics, Vol. 19, No. 5. In: 1800212 Sept./Oct. 2013

[7] Shiu DS et al. IEEE Transactions on Communications. 1999;**47**(8):1201-1210

[8] Shahine M. H., DFB Laser Diode Dynamics with Optoelectronic Feedback, D. S. Patil, Editor. Semiconductor Laser Diode Technology and Applications, IntechOpen, ISBN 978–953–51-0549-7, Chapter 6, (2012)

[9] Wan, et al., IEEE custom integrated circuits conference, pp. 25–5-1 (2005)

[10] He, et al., Analog Integrated Circuit Signal Processing Journal, Vol. 59, pp. 215–221 (2009)

[11] Xia G et al. IEEE Journal of Lightw. Technol. 2005;**23**(12):4296-4304

[12] Tang S et al. IEEE Journal of Quantum Electronics. 2001;**37**(3): 329-336

[13] Lin F et al. IEEE Journal of Quantum Electronics. 2003;**39**(4):562-568

[14] Shahine MH et al. IEEE Photonics Journal. 2010;**2**(6):1013-1026

[15] Shahine M. H., Systems and methods for real-time compensation for non-linearity in optical sources for analog signal transmission", United States Patent 7,505,496, (Mar. 17, 2009)

[16] Adler R. Proc. IRE Waves Electrons. Jun. 1946;**34**(6):351-357

[17] Helms J et al. IEEE Journal of Lightwave Technology. 1991;**9**:1567

[18] Pozar D. M., Microwave and RF Design of Wireless Systems. Hoboken, NJ: Wiley, 2001

[19] Lasri J et al. Optics Express. Jun. 2003;**11**(12):1430-1435

Chapter 2

Optoelectronics and Optical Bio-Sensors

Thamer Tabbakh, Noha Alotaibi, Zahrah A. Almusaylim, Sundos Alabdulkarim, N.Z. Jhanjhi and Nawaf Bin Darwish

Abstract

Optical biosensors (OB) have wide applications in bio-fields; they are valuable monitoring and detecting tools in therapy, food, defense and military industries. They also applied in environmental monitoring quality (i.e. water, soil and air). In recent years, biosensors have been applied in the early detection of number of diseases such as; alzahimer's disease and infecting viruses. The OB detection technology is based either on label- based or label-free method. They are composed of integral physical and biological systems, which can provide sensitive analysis for bio-analytes. This chapter will shade the light over the OB principles and their applications with the focus on the surface plasmon resonance.

Keywords: biosensors, optical , fabrication , bio-analyte, detection, monitoring

1. Introduction

OB are classified as one of the most reliable recognizing and detecting devices. Ideally, they are composed of integrated system of photonic sources and biomolecule recognition component, which in spatial contact with the transductor, that can recognize and interact with specific molecule analytes present in sample [1, 2]. The interaction induces a variation in an electrical, optical or mechanical property of the transduces, the degree of variation reflects to the quantity of the analytes [2]. The fundamental structure of OB features the 1- bio-recognition element, such as; enzymes, nucleic acids, antibodies, microbes, polysaccharides. 2- Single transductor including electrical current, potential, impedance, EM radiations, mass, optical, piezoelectric or thermal, and 3- amplifier [3–6]. The interaction occur between bio-recognition element and analyte produces a signal due to a number of factors, either alteration in proton concentration, release or capture of gases or electron, or light mission, absorption or reflectance, heat emission, change in mass [4]. The produced signal can be converted into measurable value (e.g. current, temperature alteration) by the transducer as shown in **Figure 1**.

OB provide an accurate and sensitive, timely detection technology [8]. In the current era, having such an analytical and informative features become an essence, especially in healthcare diagnostics, environment quality monitoring, food safety and security, biological warfare and biodefense [9]. Therefore, it become an attractive tool that subjected to a modification and development and emerging with nanotechnologies, microelectromechanical systems (MEMS), biotechnologies [10–12]. This reflected in the rapid advancement in biosensors, which is associated with the market growth application and is estimated by 21.18 billion USD in

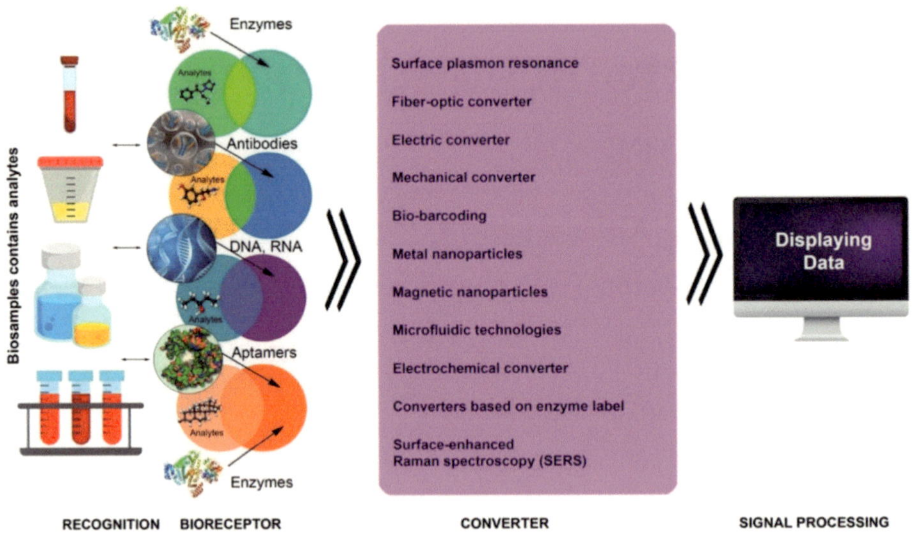

Figure 1.
Schematic diagram illustrate biosensor structure [7].

North America in the 2020 [13]. There is no universal or standard OB for detection; however, there are groups of OBs developed for specific applications and targets.

2. OB classification

There are a broad range of OB classification. Generally, the classification can be divided into two main classes: label-free and label-dependent classes.

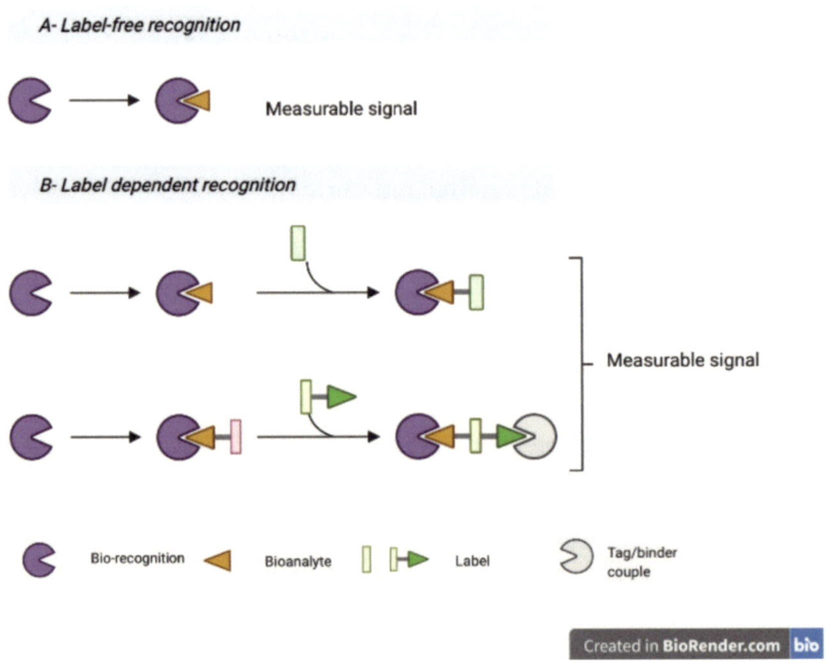

Figure 2.
Illustration graph showing label and label free recognition classes of OB.

In label-free class the measured signal is produced directly by the interaction between transducer and the bio-analyte. Whereas in label-based sensors, a label is needed to produce signal, which can be measured later by a colorimetric, luminescent or fluorescent method as can be illustrated in **Figure 2** [14].

Label-free mode is more economically efficient in comparison to label dependent mode. It requires less effort and expertise and eliminate the experimental errors such that resultant from label shelf life, signal noise background. These are commonly seen in antibody–antigen interaction using label, which has a wide application in detection technologies [15].

3. OB detector elements – Bio-recognisers

Different bio-recognisers are used in the OB detection and quantifying technologies. The diversity of the detected materials required a suitable recognizing element. Examples of analytes in the samples; genetic material, toxins, drugs, enzymes, heavy metals [16–19]. Enzymes, nucleic acids, antibodies, cells and micro-organisms are commonly used as bio-recognisers [20–24]. **Table 1** shows some of their applications and features in biosensing.

4. OB working method

An optical measurement concept is used by optical biosensors devices. Fiber optics are used along with optoelectronic transducers in these devices. The opt rode term is composed of optical and electrode terms. Enzymes and antibodies such as transducing elements are examples of what types of elements involved in these sensors. A secure non-electrical is permitted in optical biosensors, in which a sensing of equipment is inaccessible [34]. An additional advantage is that devices do not require reference sensors [35]. The reason behind that is that a light source can generate a comparative signal, which is similar to that of the sampling sensor. In order to ecxite the sensing element, optical source such as LED or Laser should be focused into substrate and photodetector capture the output signal as shown in **Figure 3** [36].

5. Types of OB

Biosensors in general divided into categories which are Bioreceptor and Transducer. While, Optical biosensors are divided into two groups, which are: direct optical biosensor detection and labeled optical biosensor detection as follow:

5.1 Evanescent wave fluorescence

Evanescent wave-based biosensors are used to investigate the exponential growth in life science applications. They include the dissociation and binding kinetics of receptor-ligand pairs and antibodies, epitope mapping, interactions between protein-DNA and DNA–DNA, phage, show libraries, and interactions between virus-protein and whole cell [37].

Waveguide interferometers have remarkable significance, because they merge both sensitive techniques that are: wave guiding and interferometry techniques. Hence, they provide great reliability and potential miniaturization and integration

Biorecognition	(Bio)sensors	Selected application	Physical transducer/liner range	Detection limit
Enzyme	Multilayers of silver metal and tantalum oxide nanoflakes with acetylcholinesterase enzyme OB [25]	Alzahimer's disease diagnosis	50–400 uM	8.709 nm/μM and a remarkable LOD value of 38 nM
	Tyrosinase on Fe3o4@Au core shell nanoparticles bio-probe [26]	Detection of dopamine, phenol and catechol	Linear response in the concentration range 5.0–75.0 μM, 10.0–100.0 μM for phenol and dopamine and 50.0–500.0 M for catechol	ND
	Carbon nanofiber gold nanoparticles tyrosinase [27]	Detection of ferulic acid in cosmetics	ND	2.89×10^{-9} mol/L
Nucleic acids	DNA - AuNTs-PC electrode	Human papilloma visus 16 & 18 [28]	0.01pM to 1 μM.	1 fM
	Disposable electrodes were fabricated by thermal evaporation on polyethylene terephthalatesubstrates covered with a nanometric gold layer manufactured in three-contact configurations	Zika virus [29]	25 nM and 340 nM	25 nM
Antibodies and Immuonsensor	Antibody against Aflatoxin immunosensor [30, 31]	Aflatoxin detection in food	ND	100 pg./mL
	Graphene oxide (GO) composite and staphylococcal protein A [32]	Human IgG detection	40–120 um/ml	10 ng/mL
Cell/Microbial -based sensor	Bioelectric Recognition Assay [33]	*L. monocytogenes* detection	ND	10^2 CFU/mL

NM = not determined.

Table 1.
Different bio-recognition, their applications and features.

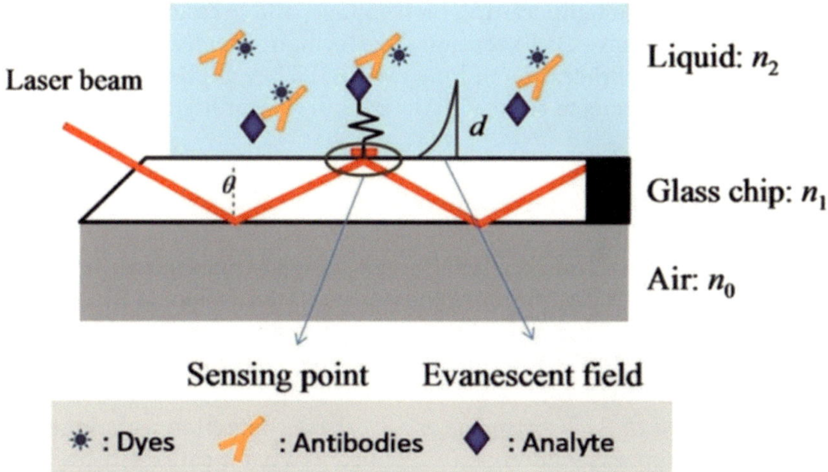

Figure 3.
Schematic for optical biosensor working method [36].

in optical chips. Through the utilization of evanescent wave technology, the interaction between biomolecules and receptors are measure by the waveguide interferometer in real time without using labels. On a sensor surface, receptors are frozen and the interaction with the close biomolecules leads to a variation in the refractive index. With being far from the surface, the evanescent wave decays exponentially, usually over the distance of 100 nm to approximately a wavelength. Due to the reason that the evanescent wave is a near-surface phenomena, therefore, by using the detection of evanescent wave stimulation to produce fluorescent signal is surface-sensitive. This means that solely fluorescent molecules close to the surface are stimulated as seen in **Figure 4** [37–40].

5.2 Optical fibers

Fiber optic is an example of analytical devices that works as a transduction item, in which it generates a signal proportional to the density of chemical or biochemical elements with react of the biological element. In addition, they transfer light with silica glass or plastic fiber optic fiber based on the Total Internal Reflection (TIR) principle to the analysis site [41]. The fiber optics biosensors are categorized into two different types:

- Intrinsic sensors: the environmental changes are transformed by the internal property of the optical fiber itself into light signal modulation. This light signal modulation may be in the form of phase, intensity, and frequency or it may be polarization [42].

- Extrinsic sensors: on the other hand, the extrinsic sensors can use the fiber as carriers of information leading to a black box. In addition, these sensors produce a light signal based on the received information at black box. This black box can be made of gas, mirrors liquid cells or several other optical signal generation mechanisms (**Figure 5**) show the difference between Intrinsic and extrinsic optical biosensor) [44].

The fiber-optical sensors essential benefits can vary from their: 1) capability of hard environment to robust EMI (electromagnetic interference immunity), chemical corrosion, high temperature, high voltage, and pressure. 2) Low power, very small size, and passive. 3) Exceptional performance such as wide bandwidth and high sensitivity. 4) Processing of long range. 5) They applied distributed or multiplexed measurements to cope with their main flaw of high cost and unfamiliarity of end-user [45].

5.3 Backscattering optical interferometric

Backscattering interferometry (BI) sensor is another category of optical biosensors. The detector can measure the uncalculated reflected intensity of a small sensing area by using a single wavelength laser light. Based on the sub wavelength formation on the top sensing area, the detector results in an interference pattern [46].

The improvement of Backscattering as a label-free detection technique appeared in field and applications as following: (a) applied to what is called lab-on-a-disc, (b) Silicon Sensor Surfaces SSS (bio reactions) application (c) Measuring minor refractive index transformations in capillaries of fused silica, and (d) Bio molecular interaction control in microfluidic channels [47].

Backscattering applications started with measuring bio molecular interactions on porous silicon based optical systems. In the pores, the surface is adjusted using elements of bio molecular recognition. Fabry-Perot fringes result in an interference pattern of impinging white light above and below the optical interference layer [48].

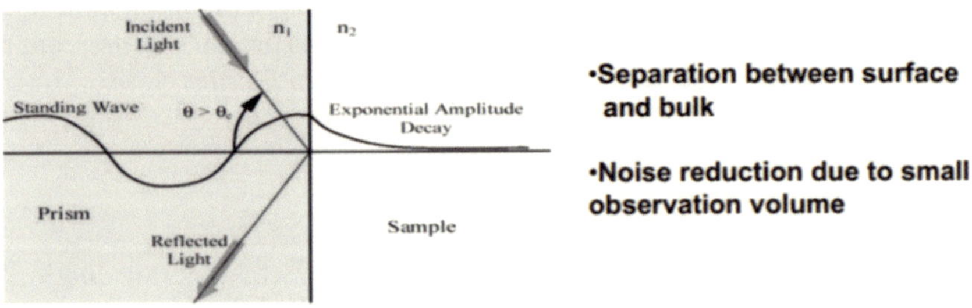

Figure 4.
Evanescent wave fluorescence biosensors working method and the separation of molecules in the surface.

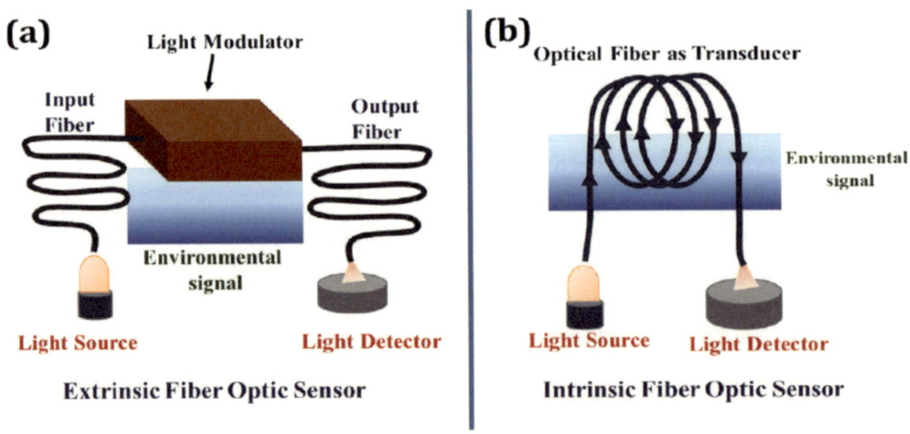

Figure 5.
(a) Extrinsic types of fiber optic sensors, and (b) intrinsic types of fiber optic sensors [43].

5.4 Reflectometric interference spectroscopy (RIfS)

In order to investigate molecular interaction, a physical technique known as reflectometric interference spectroscopy is used. This technique depends on white light being interfered at thin films as shown in **Figure 6**. In Reflectometric Interference Spectroscopy (RIfS), biomolecular reactions happen on the sensing component. The sensing component is a glass slide adjusted with a thin layer of translucent dielectric material (e.g., SiO2, SiO2–Ta2O5). When the white light strikes the reverse side of the glass, an intervention occurs from the partial beams, reflected at each interface. This intervention alternates maximum and minimum reflectance range [50], which corresponds to the constructive and destructive reflected radiation interference. Biomolecular reactions cause build-up of an adlayer on top of the dielectric, which increases the optical path length. This results in a reflectance spectrum change [51]. This change can be associated with the intensity of the reacting biomolecules and is equivalent to the increase in thickness. Information about the viscosity and refractive index of the adsorbed protein layer is given by alterations in the polarized light phase and amplitude. For the identification and quantification of diclofenac in bovine milk, this approach was used, and the detection limit obtained was 0.112 μg.

5.5 Surface-enhanced Raman scattering

Surface Enhanced Raman scattering (SERS) spectroscopy method are used for the extremely sensitive biological analytes. With rapid growth during the last four decades, surface-enhanced Raman scattering has become one of the most reliable spectroscopic method. Applications for (SERS) detection are expanding quickly in various fields such as materials science, chemistry, biochemistry, and life sciences. Remarkable growth has resulted in biological and biomedical sensing applications from advances in the creation and production of SERS-based biosensors particularly. Electromagnetic improvement leads primarily to SERS improvement, and the configurations of the hotspot are essential to the success of responsive and reproducible detection [52]. Biosensors that are SERS-based can be generated according to the sensing requirements through direct and indirect methods. To define SERS, it is an extremely sensitive optical detection method using lasers in molecules adsorbed on the top of a metal nanoparticle in order to excite vibrational transitions. The Raman cross-section for a molecule on a surface is enhanced by factors of 10 caused by large optical fields. Because of molecular vibrational events, Raman scattering depends mainly on the loss (Stokes) or gain (anti-Stokes) of energy; from inflexible scattered photons and represents the information on the molecular structure, allowing in situ and real-time detection [53, 54]. SERS is a subclass of

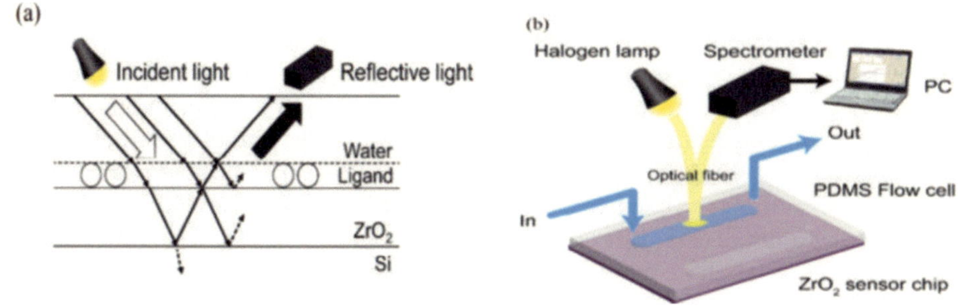

Figure 6.
Schematic illustration of (a) the RIfS principle and (b) the RIfS measurement system [49].

Raman dispersion and provides a million-fold improvement by plasmonic nanostructures, making the sensitivity of detection down to the level of a sole molecule as can be seen in **Figure** 7.

5.6 Surface plasmon resonance (SPR)

The first observation of SPR physical phenomenon was in 1902. Through decades, this observation of an esoteric optical phenomenon developed into a complete comprehension of surface plasmon physics. Then, the first successful usage of SPR was in 1983 through the fabrication of an SPR-based sensor to detect the interactions of bimolecular. Pharmacia Biosensor AB was launched the first commercial SPR-based biosensor device, which was renamed as Biacore later. Currently, several manufacturing are fabricating SPR devices. Moreover, nowadays, the SPR-based biosensor is the dominant method of biosensing [56, 57].

The SPR appears on that surface of the device, when a polarized light such as Laser or LED is illuminated to the metal surface (usually gold or silver coated service) at a particular angle and at the interface of two media (commonly water and glass). This led to the surface plasmons generation and thus a reflected light intensity reduction is created at a particular angle known as the resonance angle. This impact is proportional to the mass on the surface. To obtain a sensogram, the shift of reflectivity, wavelengths or angle are measure against time. In all configuration, label-free, direct and real-time changes of refractive index is enabled by the phenomenon of SPR at the surface of sensor, in which it is proportional to the concentration of the biomolecule as shown in **Figure 8** [58].

5.7 Liquid sensor based on optical surface plasmon resonance

With the widespread and increased demand of biological sensing devices, there has been a considered attention on reliable and multipurpose biomolecule detection systems. The motivation to produce these detection systems become greater due the rising of health awareness and spread of aging in world population. The affinity-based biosensors, which consists of a biological element and a transducer, is one of the well-known biological agent sensing devices. In the biosensor, the biological element is typically used to identify the substance that necessarily must be detected. While the transducer is used to convert the energy from one form to another, which means converting the event of bio- recognition into an electrical signal that is measurable [59, 60].

Different types of transducers for biosensors are available currently; some of them are piezoelectric transducer, optical transducer and electrochemical transducer. Optical methods have become the most know method among these transducers, which are: surface plasmon resonance (SPR) spectroscopy, interferometry, fluorescence spectroscopy and evanescent wave-based detection. In the past years, the fluorescence-based detection methods, such as Enzyme-Linked Immunosorbent Assay (ELISA), have been implemented due to their capabilities of high throughput for samples and device sensitivity. But recently the new detection methods require processing of time-consuming labeling with several procedures of protocol detection. Detection systems based on the technology of SPR based bimolecular detection have been commercialized successfully regardless of their novelty. In addition, this method simplifies real-time controlling with high sensitivity without requiring any procedures of labeling. Nevertheless, the current implemented and commercial SPR sensors are comparatively massive size systems and have low throughput, in which they limit their applications range. Hence, higher throughputs are needed

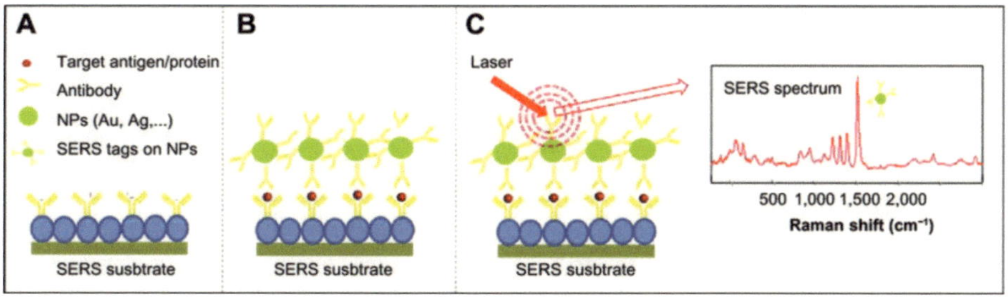

Figure 7.
(A) SERS substrate modification by antitarget antibody, (B) target isolation, followed by binding of nanoparticles (NPs), (C) labeled by SeRS tag, and SeRS-tag detection [55].

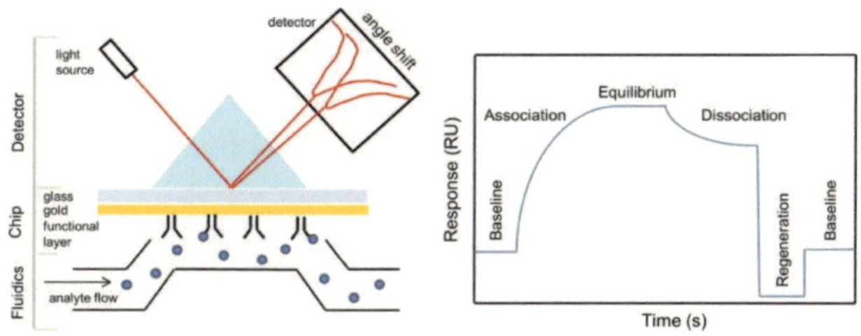

Figure 8.
The schematic of the working principle of SPR and the steps of the SPR analytical cycle.

with additional disposable and compact SPR system, even though that some of their problems have solved partially [56, 61–64].

6. SPR fabrication

This section and all the fabrication and results have been achieved previously by MQW Group at UCF [56–59, 64]. In this work, a sensor head of optical surface plasmon resonance (SPR) has illustrated in this work. It depends on an inverted-rib dielectric waveguide. The changes happen at the gold metal-dielectric interface, in which the resonance wavelength of the surface plasmon is excited. These changes are in relationship with the environment changes that occur at the top metal surface. The sensor head of the SPR with the inverted-rib dielectric waveguide composed of SU-8 polymer layer with 1.5 refractive index, whereas the cladding lower layer contains silicon oxynitride (SiOxNy) with 1.526 refractive index. The top layer is painted with a 50 nm gold thick layer. The design of sensor head of the SPR permits controlling the media of analyte with 1.44 to 1.502 refractive index. By using reference liquids collection that represent the analyte medium, an analyzer of optical spectrum and a broadband light source were utilized to measure the SPR sensor sensitivity. It was realized that when a liquid contacts the gold metal with 1.442 refractive index, the transmission spectrum has a sharp resonance dip at 1525 nm and with using a liquid of 1.502, its position was shifted to 1537 nm. Therefore, based on these measurements, the sensor devices sensitivity was specified to be S = 232 nm.RIU-1.
In this section, we demonstrate that the device can be integrated completely with a photodetection unit, a wavelength tunable light source and a liquid delivery system through microfluidic channels to make it an extremely compact unit [56–59, 64].

7. SPR simulations and design

Surface Plasmon is the Propagation of transverse magnetic (TM) surface electromagnetic waves at the dielectric interface of a gold metal. Below is the dispersion relation Eq. (1) which can be used to calculate the propagation constant.

$$k(\omega)=\frac{\omega}{c}\sqrt{\frac{\epsilon_M \epsilon_D}{\epsilon_{M+}\epsilon_D}} \tag{1}$$

The design of the reversed rib waveguide was to solely support the essential TM directed mode. The waveguide's dimensions were chosen depending on waveguide dispersion relationship in Eq. (2).

$$t_g=\frac{\operatorname{atan}\left(\sqrt{\frac{n_{s^2}-n_{c^2}}{n_{g^2}-n_{s^w}}}\right)+\pi}{k0\sqrt{n_{g^2}-n_{s^2}}}-k_0=\frac{2\pi}{\lambda}\text{ is the free}-\text{space propagation constant} \tag{2}$$

Moreover, a particular surface plasmon resonance wavelength will appear from the refractive index of the analyte material that interacts with the gold-metal. To determine theoretical dependence, matching condition criterion for the surface plasmon propagation constant (Eq. (3)) can be used.

$$\beta=\beta'+i\beta''\approx\frac{\omega}{c}\sqrt{\frac{\varepsilon'_m.\varepsilon_d}{\varepsilon'_m+\varepsilon_d}}+i\frac{\varepsilon''_m}{2(\varepsilon'_m)^2}\frac{\omega}{c}\left(\frac{\varepsilon'_m.\varepsilon_d}{\varepsilon'_m+\varepsilon_d}\right)^{3/2} \tag{3}$$

where ε_m : *Dieletric constant of metal*, ε_d : *Dieletric contant of cover material.*

After all the equation have been calculated, The SPR sensor with the propagation of the waveguide has been design and simulation as shown in **Figure 9**.

In order to define the wave vectors of the directed mode, the exact values of the refractive indices of the directing and cladding layers are important and therefore satisfy the matching condition for the SP mode to be launched. Hence, to design the top of the SPR sensor, the refractive indices for both films need to be classified carefully. To make the directing layer of the top of the SPR sensor; a SU-8 polymer with a fixed refractive index of 1.568 at 1550 nm, is used. Although the silicon oxynitride (SiOxNy) cladding layer film, generated using plasma enhanced chemical vapor deposition (PECVD), has a refractive index as shown in **Figure 10**, it can be changed by adjusting the relative nitrogen to oxygen composition of the SiOxNy film [56–59, 64].

Eq. (3) calculated the theoretical dependence as appears in **Table 2**.

The beam propagation in the SPR sensor head is shown in **Figure 11** as simulated using a RSoft beam package. It is noted here that the maximum size of the waveguide that can only support the basic TM mode are 4 μm and 500 nm respectively. The input light was given by the enhanced spontaneous emission of an erbium doped fiber transformer.

To guarantee that only the TM mode is launched to the SPR sensor waveguide by A 40× microscope objective lens, a polarizer was used. In **Figure 12**, the transmitted light at the output of the device was focused into an optical fiber. To monitor the transmission spectrum the optical fiber is connected to the Optical Spectrum Analyzer (OSA).

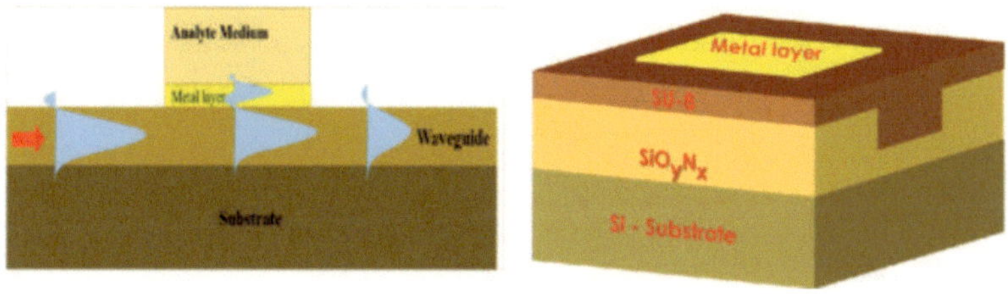

Figure 9.
Optical waveguide based surface plasmon excitation configuration and schematic of the SPR sensor head.

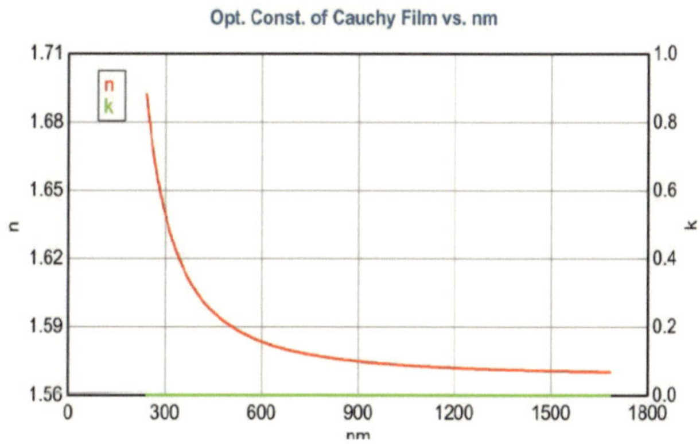

Figure 10.
Refractive index (n) and K=kelvin of SU-8 film as function of wavelength measures using ellipsometer.

Analyte medium refractive index	Surface plasmon refractive index	Theoretical wavelength (nm)
1.44	1.463	1539.5
1.47	1.484	1533.8
1.49	1.504	1530
1.5	1.515	1528.1
1.518	1.533	1524.68
1.51	1.5255	1526.2

Table 2.
Calculated wavelength vs. the surface plasmon refractive index.

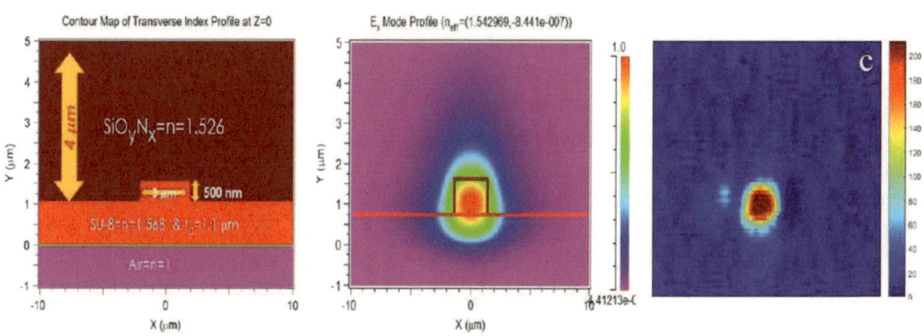

Figure 11.
(a) Simulation of SPR sensor head performed by RSoft. (b) Fundamental mode for the simulation parameter from (a). (c) Real fundamental mode after fabrication and testing.

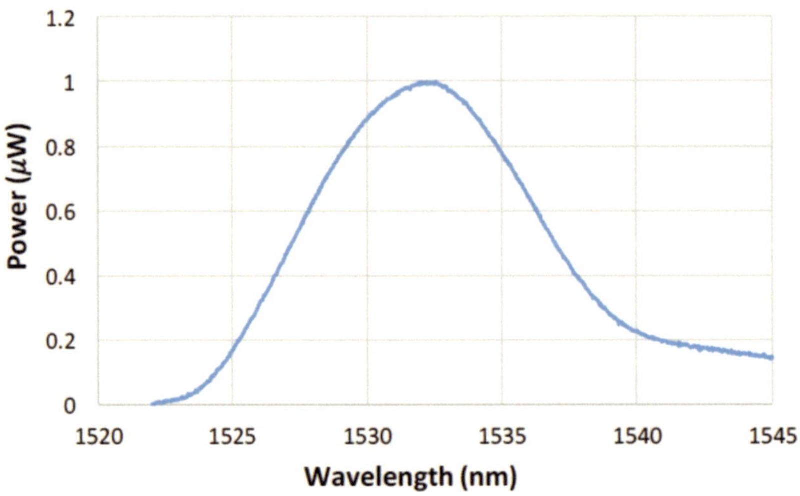

Figure 12.
The emission spectrum from the SPR sensor head before placing the sample liquids.

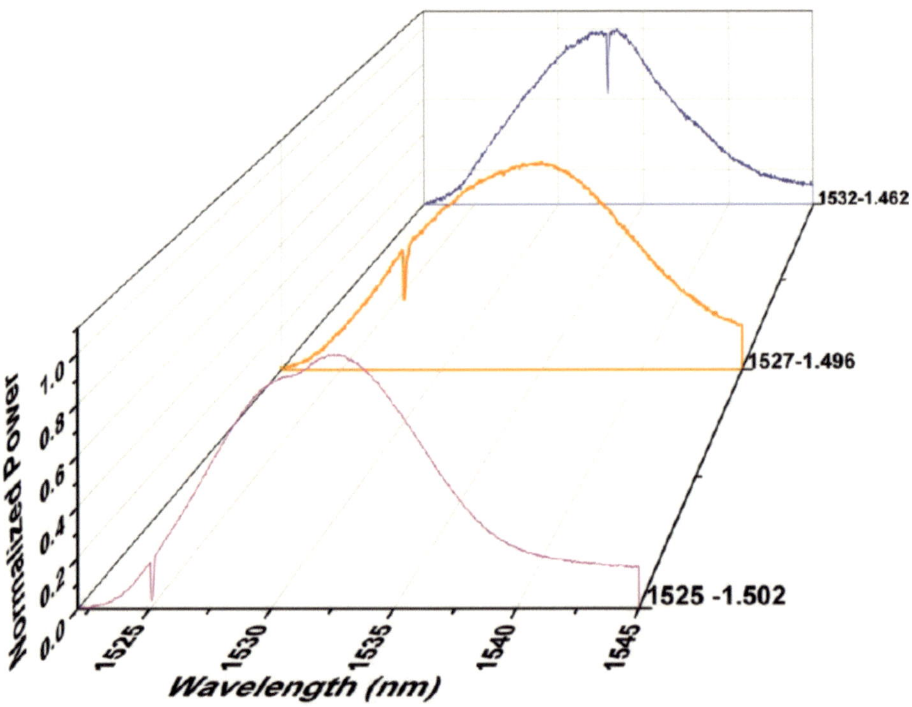

Figure 13.
Emissions spectrum after placing the different sample liquids.

It is probable that a resonant transfer of power from the waveguide can occur at the specific wavelength. The waveguide mode and the surface plasmon mode match the perfect pairing condition between them. In order for that to happen a wide spectrum of light needs to be launched into the system. For that reason, the spectrum of the transmitted light is almost the spectrum of the input light dip at the wavelength of SPR.

One after another, a set of analyte media with calibrated refractive indices (sample liquids) above the Top of the SPR sensor were implemented. To measure the transmission spectrum OSA was used to decide the wavelength of SPR. The transmission spectra for analyte media of refractive indices 1.462, 1.496, and 1.502 correspondingly. As shown in **Figure 13**, the sharp dip in the transmission spectrum moved to a shorter wavelength while lowering the sample liquid refractive index [54, 56–59].

8. Conclusion

OB detection technology offers a timely, effective and inexpensive tool to measure analytes in samples. It facilitates the detection in different areas, such as; biomedical, environmental and foods. The development and demand of using such technology will continue to grow within the current and upcoming decades, which will result in introducing new generations of OBs to the analytical and diagnostic markets worldwide.

Author details

Thamer Tabbakh[1,3], Noha Alotaibi[2,3*], Zahrah A. Almusaylim[3], Sundos Alabdulkarim[3], N.Z. Jhanjhi[4] and Nawaf Bin Darwish[5]

1 Materials Science Research Institute, King Abdulaziz City for Science and Technology, Riyadh, Kingdom of Saudi Arabia

2 Biology and Environment Research Institute, King Abdulaziz City for Science and Technology, Riyadh, Kingdom of Saudi Arabia

3 General Administration For R&D Labs, King Abdulaziz City for Science and Technology, Riyadh, Kingdom of Saudi Arabia

4 School of Computer Science and Engineering (SCE), Taylor's University, Malaysia

5 National Center for Desalination and Water Treatment Technology, King Abdulaziz City for Science and Technology (KACST), Riyadh, Saudi Arabia

*Address all correspondence to: naalotaibi@kacst.edu.sa

References

[1] Pearson JE, Gill A, Vadgama P. Analytical aspects of biosensors. Annals of Clinical Biochemistry. 2000.

[2] Thévenot DR, Toth K, Durst RA, Wilson GS. Electrochemical biosensors: Recommended definitions and classification. Biosens Bioelectron. 2001.

[3] Velasco-Garcia MN, Mottram T. Biosensor technology addressing agricultural problems. Biosystems Engineering. 2003.

[4] Turdean GL. Design and Development of Biosensors for the Detection of Heavy Metal Toxicity. Int J Electrochem. 2011.

[5] Chao J, Zhu D, Zhang Y, Wang L, Fan C. DNA nanotechnology-enabled biosensors. Biosens Bioelectron. 2016.

[6] Kahn K, Plaxco KW. Principles of biomolecular recognition. In: Recognition Receptors in Biosensors. 2010.

[7] Andryukov BG, Besednova NN, Romashko R V., Zaporozhets TS, Efimov TA. Label-free biosensors for laboratory-based diagnostics of infections: Current achievements and new trends. Biosensors. 2020.

[8] Guider R, Gandolfi D, Chalyan T, Pasquardini L, Samusenko A, Pederzolli C, et al. Sensitivity and Limit of Detection of biosensors based on ring resonators. Sens Bio-Sensing Res. 2015.

[9] Thavaselvam D, Vijayaraghavan R. Biological warfare agents. J Pharm Bioallied Sci. 2010.

[10] Fogel R, Limson J. Developing biosensors in developing countries: South Africa as a case study. Biosensors. 2016.

[11] Dey D, Goswami T. Optical biosensors: A revolution towards quantum nanoscale electronics device fabrication. Journal of Biomedicine and Biotechnology. 2011.

[12] Chen JC, Chou JC, Sun TP, Hsiung SK. Portable urea biosensor based on the extended-gate field effect transistor. Sensors Actuators, B Chem. 2003.

[13] Indahningrum R putri. Biosensors Market Size, Share & Trends Analysis Report By Application (Medical, Agriculture, Bioreactor) By Technology (Thermal, Electrochemical, Optical), By End Use, By Region, And Segment Forecasts, 2020-2027. Vol. 2507, Grand view research. 2020.

[14] Damborský P, Švitel J, Katrlík J. Optical biosensors. Essays Biochem. 2016.

[15] Kusnezow W, Jacob A, Walijew A, Diehl F, Hoheisel JD. Antibody microarrays: An evaluation of production parameters. Proteomics. 2003.

[16] Thevenot D, Toth K, Durst R, Wilson G, Thevenot D, Toth K, et al. Electrochemical biosensors : Recommended definitions and classification (Technical Report) To cite this version : ELECTROCHEMICAL BIOSENSORS : RECOMMENDED. Elsevier. 2001;121-131.

[17] Mohanty SP, Koucianos E. Biosensors: A tutorial review. IEEE Potentials. 2006;25(2):35-40.

[18] Turner APF. Biosensors - Sense and sensitivity. Science. 2000.

[19] Chen H, Shao S, Yu Y, Huang Y, Zhu X, Zhang S, et al. A dual-responsive biosensor for blood lead detection. Anal Chim Acta. 2020.

[20] Yemini M, Reches M, Gazit E, Rishpon J. Peptide nanotube-modified electrodes for enzyme-biosensor applications. Anal Chem. 2005.

[21] Wang X, Niu S, Wei M, Liu S, Liu R, Shi C, et al. Ultrasensitive electrochemical DNA biosensor based on a tetrahedral structure and proximity-dependent surface hybridization. Analyst. 2020.

[22] Aydin EB, Aydin M, Sezgintürk MK. The development of an ultra-sensitive electrochemical immunosensor using a PPyr-NHS functionalized disposable ITO sheet for the detection of interleukin 6 in real human serums. New J Chem. 2020.

[23] Kharkova AS, Arlyapov VA, Turovskaya AD, Shvets VI, Reshetilov AN. A mediator microbial biosensor for assaying general toxicity. Enzyme Microb Technol. 2020.

[24] Hadjilouka A, Loizou K, Apostolou T, Dougiakis L, Inglezakis A, Tsaltas D. A Cell-Based Biosensor System for Listeria monocytogenes Detection in Food. Proceedings. 2020;60(1):49.

[25] Kant R, Gupta BD. Fiber-Optic SPR Based Acetylcholine Biosensor Using Enzyme Functionalized Ta2O5 Nanoflakes for Alzheimer's Disease Diagnosis. J Light Technol. 2018.

[26] Arkan E, Karami C, Rafipur R. Immobilization of tyrosinase on Fe3o4@ Au core–shell nanoparticles as bio-probe for detection of dopamine, phenol and catechol. J Biol Inorg Chem. 2019.

[27] Bounegru AV, Apetrei C. Development of a novel electrochemical biosensor based on carbon nanofibers–gold nanoparticles– tyrosinase for the detection of ferulic acid in cosmetics. Sensors (Switzerland). 2020.

[28] Shariati M, Ghorbani M, Sasanpour P, Karimizefreh A. An ultrasensitive label free human papilloma virus DNA biosensor using gold nanotubes based on nanoporous polycarbonate in electrical alignment. Anal Chim Acta. 2019.

[29] Faria HAM, Zucolotto V. Label-free electrochemical DNA biosensor for zika virus identification. Biosens Bioelectron. 2019.

[30] Kim HJ, Choi W, Kim J, Choi J, Choi N, Hwang KS. Highly sensitive three-dimensional interdigitated microelectrode biosensors embedded with porosity tunable hydrogel for detecting proteins. Sensors Actuators, B Chem. 2020.

[31] An Z, Jang CH. Label-free optical detection of aflatoxin by using a liquid crystal-based immunosensor. Microchem J. 2018.

[32] Wang Q, Wang B. Sensitivity enhanced SPR immunosensor based on graphene oxide and SPA co-modified photonic crystal fiber. Opt Laser Technol. 2018.

[33] Hadjilouka A, Loizou K, Apostolou T, Dougiakis L, Inglezakis A, Tsaltas D. Newly Developed System for the Robust Detection of *Listeria monocytogenes* Based on a Bioelectric Cell Biosensor. Biosensors. 2020.

[34] Mohanty, S.P. and E. Kougianos, Biosensors: a tutorial review. Ieee Potentials, 2006. 25(2): p. 35-40.

[35] Michenfelder, G., et al., Optical sensor. 2002, Google Patents.

[36] Xiao-Hong, Z., et al., A reusable evanescent wave immunosensor for highly sensitive detection of bisphenol A in water samples. Scientific reports, 2014. 4: p. 4572.

[37] Taitt, C.R., G.P. Anderson, and F.S. Ligler, Evanescent wave fluorescence biosensors. Biosensors and

Bioelectronics, 2005. 20(12): p. 2470-2487.

[38] Taitt, C.R., G.P. Anderson, and F.S. Ligler, Evanescent wave fluorescence biosensors: Advances of the last decade. Biosensors and Bioelectronics, 2016. 76: p. 103-112.

[39] Blair, S. and Y. Chen, Resonant-enhanced evanescent-wave fluorescence biosensing with cylindrical optical cavities. Applied Optics, 2001. 40(4): p. 570-582.

[40] Golden, J.P., et al., An evanescent wave biosensor. II. Fluorescent signal acquisition from tapered fiber optic probes. IEEE Transactions on Biomedical Engineering, 1994. 41(6): p. 585-591.

[41] Udd, E., An overview of fiber-optic sensors. review of scientific instruments, 1995. 66(8): p. 4015-4030.

[42] Fidanboylu, K. and H. Efendioglu. Fiber optic sensors and their applications. in 5th International Advanced Technologies Symposium (IATS'09). 2009.

[43] Méndez, A. and A. Csipkes, Overview of fiber optic sensors for NDT applications, in Nondestructive testing of materials and structures. 2013, Springer. p. 179-184.

[44] Udd, E., Overview of fiber optic sensors. Fiber Optic Sensors, 2002: p. 1-40.

[45] Zhao, Y., et al., Applications of fiber-optic biochemical sensor in microfluidic chips: a review. Biosensors and Bioelectronics, 2020: p. 112447.

[46] Khan, M. and S.-Y. Park, Liquid crystal-based biosensor with backscattering interferometry: a quantitative approach. Biosensors and Bioelectronics, 2017. 87: p. 976-983.

[47] Werquin, S., S. Verstuyft, and P. Bienstman, Integrated interferometric approach to solve microring resonance splitting in biosensor applications. Optics express, 2013. 21(14): p. 16955-16963.

[48] Sørensen, H.S., et al., Biosensing with backscattering interferometry. Biomedical Optics Medical Imaging IN SPIE Newsroom, 2009. 5.

[49] Kurihara, Y., T. Sawazumi, and T. Takeuchi, Exploration of interactions between membrane proteins embedded in supported lipid bilayers and their antibodies by reflectometric interference spectroscopy-based sensing. Analyst, 2014. 139(22): p. 6016-6021.

[50] Proll, G., et al., Reflectometric interference spectroscopy, in Biosensors and Biodetection. 2017, Springer. p. 207-220.

[51] Proll, G., et al., Reflectometric interference spectroscopy, in Biosensors and Biodetection. 2009, Springer. p. 167-178.

[52] Quang, L.X., et al., A portable surface-enhanced Raman scattering sensor integrated with a lab-on-a-chip for field analysis. Lab on a Chip, 2008. 8(12): p. 2214-2219.

[53] Zhang, Y., et al., Surface-enhanced Raman scattering sensor based on D-shaped fiber. Applied Physics Letters, 2005. 87(12): p. 123105.

[54] Xie, Y., et al., A novel surface-enhanced Raman scattering sensor to detect prohibited colorants in food by graphene/silver nanocomposite. Talanta, 2012. 100: p. 32-37.

[55] Krejcova, L., et al., Nanoscale virus biosensors: state of the art. Nanobiosensors in Disease Diagnosis, 2015. 4: p. 47-66.

[56] Tabbakh, T.A. and P.L. Likamwa, Monolithically integrated InGaAsP multiple quantum well tunable laser diode for integrated optic surface plasmon resonance sensing. Optical Engineering, 2018. 57(12): p. 120503.

[57] Tabbakh, T. and P. LiKamWa. Intermixed InGaAsP MQW tunable laser diode suitable for probing surface plasmon resonance optical sensor. in Nanoengineering: Fabrication, Properties, Optics, and Devices XV. 2018. International Society for Optics and Photonics.

[58] Tabbakh, T., Monolithically Integrated Wavelength Tunable Laser Diode for Integrated Optic Surface Plasmon Resonance Sensing. 2018.

[59] Tabbakh, T. and P. LiKamWa. Liquid sensor based on optical surface plasmon resonance in a dielectric waveguide. in Micro-and Nanotechnology Sensors, Systems, and Applications X. 2018. International Society for Optics and Photonics.

[60] Batchelder, D.N. and J.P. Willson, Optical surface plasmon sensor device. 1989, Google Patents.

[61] Steiner, G., Surface plasmon resonance imaging. Analytical and bioanalytical chemistry, 2004. 379(3): p. 328-331.

[62] Bokken, G.C., et al., Immunochemical detection of Salmonella group B, D and E using an optical surface plasmon resonance biosensor. FEMS microbiology letters, 2003. 222(1): p. 75-82.

[63] Aleahmad, P., et al. Controllable red and blue bandgap energy shifted LEDs and modulators on InGaAsP quantum well platform. in Integrated Optics: Devices, Materials, and Technologies XX. 2016. International Society for Optics and Photonics.

[64] Tabbakh, T. and P. LiKamWa. Quantum well intermixed tunable wavelength single stripe laser diode. in Active Photonic Platforms IX. 2017. International Society for Optics and Photonics.

Chapter 3

Piezo-Optical Transducers in High Sensitive Strain Measurements

Andrey G. Paulish, Peter S. Zagubisalo, Sergey M. Churilov, Vladimir N. Barakov, Mikhail A. Pavlov and Alexander V. Poyarkov

Abstract

New piezo-optical sensors based on the piezo-optical effect for high sensitive mechanical stress measurements have been proposed and developed. The piezo-optical method provides the highest sensitivity to strains compared to sensors based on any other physical principles. Piezo-optical sensors use materials whose parameters practically not change under load or over time, therefore piezo-optical sensors are devoid of the disadvantages inherent in strain-resistive and piezoelectric sensors, such as hysteresis, parameters degradation with time, small dynamic range, low sensitivity to strains, and high sensitivity to overloads. Accurate numerical simulation and experimental investigations of the piezo-optical transducer output signal formation made it possible to optimize its design and show that the its gauge factor is two to three orders of magnitude higher than the gauge factors of sensors of other types. The cruciform shape of the transducer photoelastic element made it possible to significantly increase the stresses in its working area at a given external force. Combining compactness, reliability, resistance to overloads, linearity and high sensitivity, in terms of the all set of these parameters, piezo-optical sensors significantly surpass the currently widely used strain-resistive, piezoelectric and fiber-optic sensors and open up new, previously inaccessible, possibilities in the tasks of measuring power loads.

Keywords: piezo-optical transducers, strain gauge, sensor gauge factor, photoelasticity, optoelectronic devices

1. Introduction

Optoelectronic measurement methods are based on optical effects associated with the electromagnetic radiation interaction with matter. The polarization of the electromagnetic wave during such interaction is the most "susceptible" parameter which ensures high sensitivity of polarization-optical methods in comparison with other measurement methods [1–5]. In addition, the optical measurement method is free from electromagnetic interference and can be used in severe environmental conditions and at high temperature [6].

One of the most important directions in the development of measuring methods and sensors based on them is the monitoring of stress states in various structures both in industry and research-and-development activities. Modern and promising strain sensors should have low weight, small size, low power, resistance to environmental influences and electromagnetic noise immunity, stability of parameters during operation, and low cost. Today, the most widely used method for strain measuring is based on the strain-resistive effect. The strain-resistive devices are used due to its relatively low cost and easy-to-use design [7]. However, such sensors have a number of unavoidable drawbacks: parameters degradation with time, hysteresis, nonlinearity, small dynamic range, low deformation sensitivity, and dramatic sensitivity to the overloads [7]. However, with the development of technologies, especially precision ones, the requirements for strain gauges increase significantly and strain-resistive gauges do not meet modern requirements.

For measuring vibrations, accelerations, acoustic signals, sensors based on the piezoelectric effect are widely used [8–12]. Such sensors performed well when measuring dynamic deformations (vibrations), but they are not suitable for measuring static loads due to the leakage of the charge induced by the load. Moreover, when such sensors are operated, both reversible and irreversible changes in their gauge factor and other characteristics are possible. This naturally limits the application conditions and is one of the most serious drawbacks of piezoelectric accelerometers.

Fiber-optic sensors are among the modern optical methods for measuring strain. A significant advantage of such sensors is the ability to implement several, up to a hundred, sensors on single optical fiber, which is used in distributed monitoring systems [13–19]. The disadvantages of such sensors include, first of all, low sensitivity (lower than that of strain-resistive gages and piezoelectric ones) and a complex system of optical measurements.

The most sensitive method for strain measuring is based on the piezo-optical effect, which consists in changing the polarization of light propagating in a transparent stressed material [7, 20]. Such sensors have a significantly higher sensitivity than strain-resistive ones due to the fundamentally high light polarization sensitivity to change in the state of the substance in which light propagates [20]. The attempts to develop the industrially usable deformation sensors based on the piezo-optical measuring transducers are known in the literature [21–24]. However, for a number of reasons, primarily of a technological nature, these developments did not lead to the appearance of piezo-optical strain gauges capable of competing with strain-resistive sensors in terms of a price/quality ratio. In the scientific literature, there are no data on the comparative analysis of the gauge factors (the main strain gauge parameter) of strain sensors based on different physical principles, which complicates the objective assessment of their advantages and disadvantages.

The purpose of this work is to develop theoretical foundations and basic design and technological solutions for creating a highly sensitive strain sensor based on a piezo-optical optoelectronic transducer. At the same time, the sensor must meet the requirements of industrial operation, significantly surpass the parameters of modern sensors based on other principles, and be devoid of drawbacks inherent in these sensors: parameters degradation with time, hysteresis, nonlinearity, small dynamic range, low sensitivity to the deformation, and high sensitivity to the overloads.

2. Comparative analysis of the sensitivity of strain gauges based on various physical principles

Here we consider the theoretical foundations of the physical effects that underlie modern strain gauges. We will also determine the gauge factor (GF) for each type of sensor so that the sensitivity of these sensors can be compared. The GF dependence on the sensor design will also be determined.

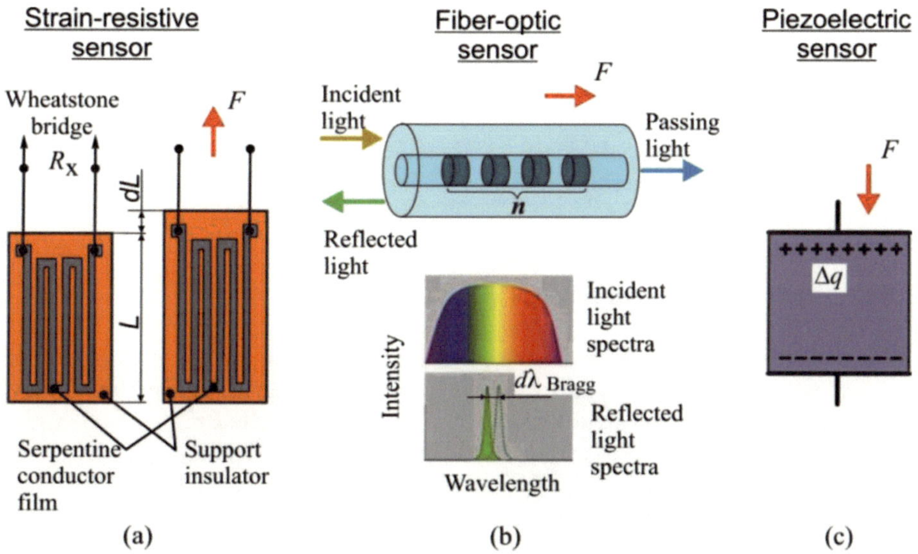

Figure 1.
Operation principle of strain gauges based on: (a) Strain-resistive, (b) Fiber–optic and (c) Piezoelectric sensors. F – Measured load, R_x – Alternating resistance, n – Effective refractive index, Δq – Stress-induced charge density.

2.1 Strain-resistive effect and strain gauge sensors

Strain-resistive gauge sensors have been well known for a long time. At first glance, the design of such a sensor looks rather simple (**Figure 1a**). A typical strain sensor design is a thin serpentine conductor film (thickness – around 0.0025 mm and length – 0.2–150 mm) deposited on a thin polymer support film. The conductor film thickness is made to be thinner and the length longer to obtain a sufficiently large resistance. Therefore, the resistance creates sufficiently large voltage for the measurements. The structure is glued to a controlled specimen and incorporated into the Wheatstone bridge [7] as an alternating resistance R_x (**Figure 1a**). The polymer film is the carrier and insulator. Their production technology is well developed and well controlled. Nevertheless, polymer films, glues, and thin metal films have low plastic deformation threshold. This leads to the problems listed above: hysteresis, nonlinearity, degradation of parameters with time, etc.

The theoretical foundations of such sensors operation, as well as the technical aspects of their use, are described in detail in work [7]. Here, the basic provisions necessary for comparing such sensors with sensors based on other physical principles are briefly presented.

It is well known that the conductor length increases $(L + dL)$, and its cross-section decreases $(S - dS)$ under the action of force F along the conductor (**Figure 1a**). In case of elastic deformation, the change in relative resistance dR/R is given by expression [7]:

$$\frac{dR}{R} = \frac{d\rho}{\rho} + \frac{dL}{L} + 2\nu\frac{dL}{L} - \nu^2\left(\frac{dL}{L}\right)^2, \tag{1}$$

where ρ is the specific resistivity and ν is the Poisson ratio, usually equal to 0.3 [7]. The strain sensitivity, gauge factor, connects the relative deformation value (dL/L) with the relative change in the measuring parameter (signal). For the strain-resistive sensor the GF is determined by the expression [7]:

$$GF = \frac{dR/R}{dL/L} = \frac{d\rho/\rho}{dL/L} + 1 + 2\nu - \nu^2\left(\frac{dL}{L}\right). \quad (2)$$

It can be seen from Eq. (2) that the GF magnitude depends only on the properties of the conductor material ($d\rho/\rho$) and is *independent* of the strain gauge design. Typical gauge factor values for the materials used to create the sensors lie in the range of 2–6 [7].

2.2 Fiber-optic strain gauge sensors

The sensitive element of the fiber-optic sensors is a Bragg fiber grating which is an optical fiber with a periodically changing refractive index (**Figure 1b**). When light passes through such a structure, part of it with a specific Bragg wavelength (λ_B) is reflected, and the rest is transmitted further. The reflected light wavelength λ_B is given by the relation $\lambda_B = 2nL$, where n is the effective refractive index of the optical fiber and L is the distance between the gratings or the grating period. Stretching/compression of the fiber changes the distance L and the refractive index n resulting in a change in Bragg wavelength ($\lambda_B \pm d\lambda_B$), which is recorded by the optical system. The fiber-optic strain sensor GF is found by analogy with strain-resistive sensors [25]:

$$GF = \frac{d\lambda_B/\lambda_B}{dL/L}, \quad (3)$$

where $d\lambda_B$ is the change in the reflected-light wavelength during deformation of the fiber grating and dL/L is the relative deformation of the grating. According to [25], the fiber-optic strain sensor GF is about 0.78, which is markedly lower than the strain-resistive sensor GFs. As in case of strain-resistive sensors, it does not depend on the sensor design and is determined by the properties of the fiber-optic material.

2.3 Piezoelectric effect and piezoelectric transducers

A piezoelectric transducer converts a mechanical force into an electric charge. Its operation is based on the piezoelectric effect which entails the occurrence of dielectric polarization under mechanical stresses (**Figure 1c**). The density of the electric charge induced on the piezoelectric element surface under an external load is described by the Equation [26].

$$\Delta q = d_{ij}\sigma, \quad (4)$$

where Δq is the surface charge density; d_{ij} is the piezoelectric modulus described by a 3×6 matrix with typical component values in the range of 10^{-10}–10^{-12} C/N; σ is the stress in the material under the external load. The sensitivity of these sensors is described by the piezoelectric modulus which complicates their comparison with strain-resistive and fiber-optic sensors. Similarly to Eqs. (2) and (3), the piezoelectric sensor GF should be inversely proportional to the relative deformation dL/L. Using Hooke's law $\sigma = E(dL/L)$ and Eq. (4) we get:

$$GF = \frac{\Delta q}{dL/L} = \frac{d_{ij}\sigma}{dL/L} = \frac{d_{ij}E(dL/L)}{dL/L} = d_{ij}E. \quad (5)$$

It follows from Eq. (5) that the GF does not depend on the sensor design and is determined only by the piezoelectric element material properties. **Table 1** shows the values of the piezoelectric moduli and Young's moduli taken from [26, 27] and the sensitivity factors calculated by Eq. (5) for some materials widely used for the fabrication of piezoelectric sensors. It is the GF (and not the piezoelectric modulus) that is an objective sensitivity parameter of piezoelectric sensors when compared with the sensitivity of another type of sensors. For example, the piezoelectric modulus for electroactive polymers is more than two orders of magnitude greater than for other piezomaterials. However, this advantage almost vanishes due to the small elastic modulus, and, as a result, GF becomes two orders of magnitude lower than that of the other materials. **Table 1** shows that the piezoelectric sensor GFs are comparable in order of magnitude with the strain-resistive sensor GFs.

2.4 Piezo-optical effect and piezo-optical transducers

The piezo-optical effect (also called "photoelasticity") used for precision stress (deformation) measurements is known since the 1930s [28]. If a light wave with a linear (circular) polarization (**Figure 2**) is incident upon transparent material

Material	d_{ij}, 10^{-12} C/N	E,GPa	GF
Lead zirconate titanate (PZT)			
PZT −19	160–330	70	11,2–23,1
PZT −21	40–100	90	2,8–7,0
PZTNV-1	160–400	64	10,2–26,6
PZT-5A	274–593	60	16,4–35,6
PZT-6A	80–189	94	7,5–17,8
PZT-6B	27–71	111	3,0–7,8
Crystal quartz	2,33	78,7	0,16
Barium titanate (BaTiO3)	78	100	7,8
Solid solutions (K, Na)NbO3	80–160	104–123	8,3–19,7
Nonpolar polymers (polyethylene, rubbers, etc.)	0,1–1,0	1–3	$0{,}1–3{\cdot}10^{-3}$
Polar polymers (polycarbonate, polyvinyl chloride)	20–40	2,3–3,5	0,05–0,14
Electroactive polymers	30,000	$(6–1000){\cdot}10^{-6}$	< 0,03

Table 1.
Parameters of some piezoelectric materials.

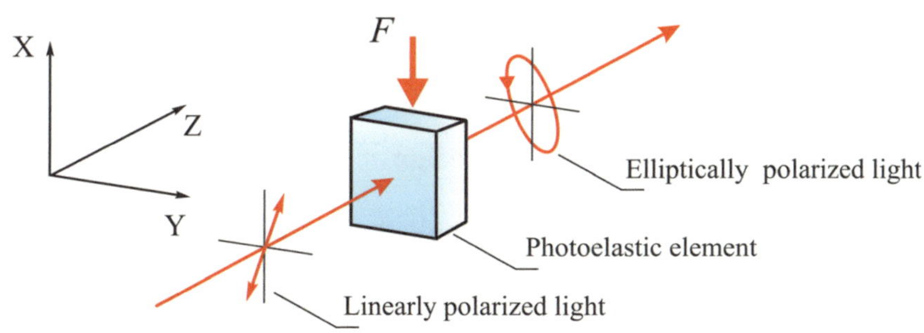

Figure 2.
Effect of photoelasticity.

(photoelastic element) under the load F, an additional phase difference Δ between the polarization components, perpendicular and parallel to the stress axis, arises due to the double refraction [20]. As a result, the light polarization in the general case becomes elliptical. The magnitude of the phase difference Δ is determined by the expression

$$\Delta = \frac{2\pi d}{\lambda}(n_o - n_e) = \frac{2\pi d}{\lambda}K(\sigma_y - \sigma_x) = \frac{2\pi d}{\lambda}KE\varepsilon, \tag{6}$$

here d is the path length of a light beam in the stressed material (photoelastic element thickness in the piezo-optical transducer). λ – working wavelength. n_o and n_e – refractive indexes for ordinary and extraordinary rays. K – stress-optical coefficient with typical value 10^{-11}–10^{-12} m^2/N. σ_x and σ_y – tensions along and across the applied force in a plane perpendicular to the direction of light propagation. E – Young modulus of the optical material. $\varepsilon = dL/L$ – relative deformation of the optical material. In a general case, the stresses in a photoelastic element are described by the Cauchy stress tensor σ_{ij}. Eq. (6) describes the effect of photoelasticity (**Figure 2**).

The GF for the piezo-optical transducer can be determined by analogy with the strain-resistive one, where the measuring parameter is dU/U, which is equivalent to the dR/R at constant current (Eq. (2)).

The measuring parameter for the piezo-optical effect is the phase difference Δ (Eq. (6)), which is measured by the ellipsometry techniques, so the expression for the piezo-optical GF takes the form:

$$\text{GF} = \frac{\Delta}{\varepsilon} = \frac{2\pi d}{\lambda}KE. \tag{7}$$

The GF magnitude depends *not only* on the material constants (K, E) but also on the design of the piezo-optical transducer (d and λ) (see Eq. (7)). In addition, the stresses magnitude $(\sigma_y - \sigma_x)$ in the photoelastic element depends strongly on the photoelastic element design to which the force is applied. This opens up the possibilities for optimizing the piezo-optical transducer parameters to increase its sensitivity to the applied force.

If fused quartz is used as the optical material, the gauge factor value GF_{theor} can be calculated according to Eq. (7) taking into account $K = 3.5 \times 10^{-12}$ m^2/N [29], $E = 70$ GPa, photoelastic element thickness $d = 4$ mm and $\lambda = 660$ nm at 20°C (conventional LED):

$$\text{GF}_{\text{theor}} = \frac{2\pi \cdot 4 \cdot 10^{-3}}{0.66 \cdot 10^{-6}} 3.5 \cdot 10^{-12} \cdot 7 \cdot 10^{10} = 9330. \tag{8}$$

The GF value is more than *three orders* of magnitude higher than the strain-resistive effect values [30].

3. Piezo-optical transducer of new design

In order to achieve the set goal of the work, the following was done [30].

1. We have studied the process of piezo-optical transducer output signal generating in detail with the help of accurate numerical simulation. We determined the piezo-optical sensor GF and compared it with other types [30, 31].

2. We showed that cruciform photoelastic element (PE) allows us to significantly increase the stresses magnitude in the PE working area under the external force action and, thereby, increase the sensitivity to the force [32].

3. Fused quartz [33] was chosen as the photoelastic element material although it has a smaller stress-optical coefficient K compared to any crystals or solid polymers. However, fused quartz technology is inexpensive and well-developed. No plastic deformation exists in fused quartz and its elastic properties do not change with time. It offers a high compression damage threshold, thus, ensuring an overload resistance and a wide dynamic range of deformation measurement. Thus, there are no hysteresis and parameters degradation in such strain gauges.

4. Due to the cruciform PE, the remaining optical elements can be located within the PE dimension, and, consequently, the transducer can have its own unified casing and its technology is separated from the load cell technology. The attachment to the monitored object is carried out with the help of load elements, the design of which depends on the mounting method [34].

5. In the mounted state, the PE is under the preliminary compressive stresses along two orthogonal directions [35], which ensures: i) a reliable glueless force-closure between the PE and the load element; ii) the operation both in compression and in tension; iii) the temperature independence of the device output signal since temperature changes do not change the pre-stressed isotropy.

6. The output signal electrical circuit is located inside the transducer shielded housing and has any desired interfaces. As a result, the transducer is a complete device that does not require a secondary signal transducer as is the case with strain-resistive sensors [36].

As a result, we were able to optimize the transducer design and significantly reduce the production cost while maintaining high field-performance data. To confirm this, we compared its parameters with the parameters of most sensitive strain-resistive sensor used to calibrate the deadweight machines (see Section 6).

The optical scheme of the piezo-optical transducer is shown in **Figure 3** and consists of an optically coupled light-emitting diode (LED), a polarizer (Pol), a quarter-wave plate ($\lambda/4$), a photoelastic element (PE), two analyzers (An1, An2) and two photodetectors (PD1, PD2) [30]. The measured force F is applied to the PE.

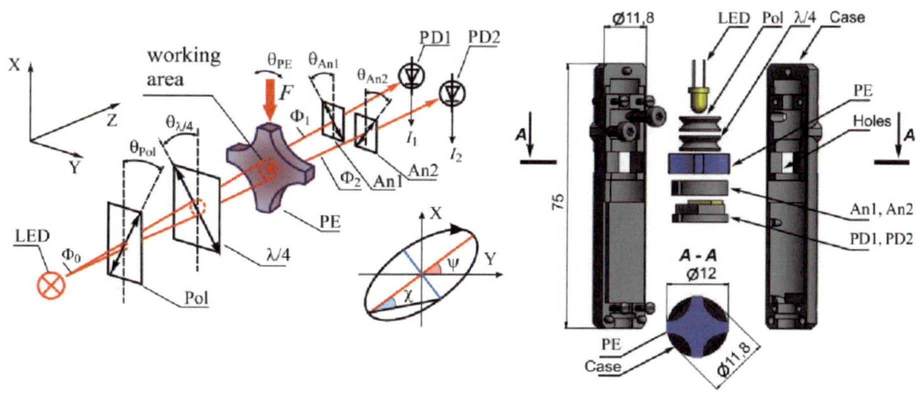

Figure 3.
Optical scheme of piezo-optical transducer (left) and its design (right).

The analyzers axes are oriented at the angle of 90°. The photoelastic element is the main component of the piezo-optical transducer. The working area of the PE is limited by the part (dashed circle in **Figure 3**) passing through which the light rays hit the photosensitive areas of the photodetectors. The rest of the PE does not participate in photodetectors signals. A feature of the transducer's optical scheme is the separation of the light beam along the front of the incident wave into two beams before falling onto the photoelastic element. This solution allows the use of film polarizers (Polaroid) reducing the optical path of light beams and, consequently, the dimensions of the converter and also allowing the use of an incoherent light source with low power consumption. The size of the optical scheme does not exceed a cubic centimeter.

The phase difference Δ between two orthogonal components of the beam polarization caused by stresses in the PE working area leads to the change in light beams intensities (Φ_1, Φ_2 in **Figure 3** on the left) incident on the photodetectors. Accordingly, it leads to the change in the output of electrical signals (I_1, I_2 in **Figure 3** on the left). The transducer electronic circuit generates a differential output signal $\sim (I_1 - I_2)/(I_1 + I_2) \sim \Delta$ which is proportional to the $\Delta\sigma = \sigma_y - \sigma_x$ (Eq. (6)) and to the applied load value.

We have optimized the PE shape and showed that the cruciform PE allows us to significantly increase the stresses magnitude in the PE working area for a given applied force [32]. The results of numerical modeling for stresses in PEs of various shapes (square, circular, rhombic, and cruciform) subjected to the same external force F = 4 N are shown in **Figure 4** [30–32]. The $\Delta\sigma = \sigma_y - \sigma_x$ isolines show the stresses distribution in PEs ($\Delta\sigma$ magnitudes are expressed in MPa). The PE working area which light passes when falling on photodetectors is shown with dashed circles. The overall dimension of all PEs was 12 × 12 mm. The calculated points are connected by straight lines just for convenience. It can be seen that the stresses are concentrated near the force application points and they are reducing considerably toward the PE center. Thus, the PE working area falls into the PE part where the stresses are minimal.

The transition to the cruciform PE and the increase in "dent" depth h (**Figure 4**) result in the stresses redistribution toward the PE center and the increase in the stress in the PE working area. For the PE shape shown in **Figure 4a** (bottom right), the increase in stresses averaged over the PE working area was 2.1 times higher compared to the square and circular shapes (dependence (1) in **Figure 4b**).

However, it is evident that the mechanical strength of a PE should go down as the "dent" h gets deeper. This issue was investigated by calculating the PE damage threshold for various PE shapes. The magnitude of stress arising in PEs of various shapes was

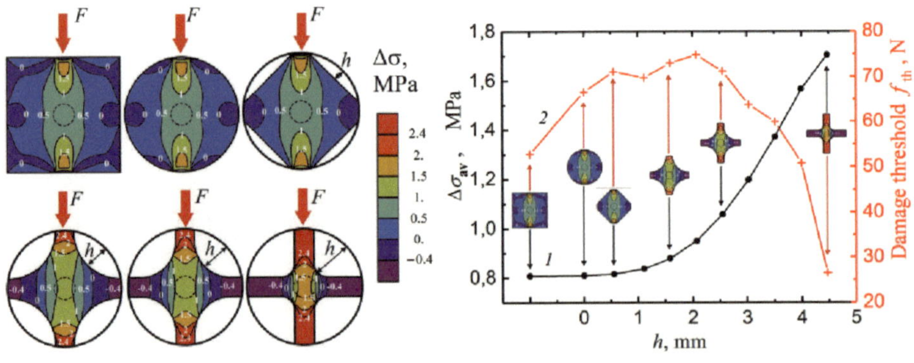

Figure 4.
(a) Isolines of stress difference $\Delta\sigma$ for the photoelastic elements of different shapes; (b) The dependences of the stress difference $\Delta\sigma_{av}$, averaged over the PE working area (1), and damage threshold f_{th} (2) on depth h [32].

compared to the ultimate strength of quartz (51.7 MPa) [37]. The dependence (2) in **Figure 4b** shows the threshold force f_{th} under which the PE breakdown occurs. It can be seen that the damage threshold increases with a change in PE form from square to circular then to rhombus and even continues to grow with an increase in the "dent" depth h up to 2 mm. Further increase in h resulted in a reduction in f_{th}. The threshold begins to decrease rapidly only when $h > 3.5$ mm. Thus, at the same damage threshold for $h \approx 3.5$ mm, we have a significant increase in stresses in the PE working area.

4. Piezo-optical transducer model for numerical simulation

The mathematical models of the transducer were created for the accurate numerical simulation of its operation. The first model is for the simulation of the light parameters transformation as it passes through the optical elements of the piezo-optical transducer. The second model is for the simulation of stresses spatial distribution in the PE body and its deformation under applied force. The models, the equations used are described in detail in [30]. The initial data and the main results obtained are briefly presented here.

4.1 Simulation of the light propagation

The optical scheme of the piezo-optical transducer showed in **Figure 3** on the left was used for the simulations. The simulations were performed using the Wolfram Mathematica™ package and took into account the design of the developed transducer: the radiation diagram of LEDs used, the dimensions of the photosensitive areas of photodiodes used, light refraction in the PE, the distances between the transducer elements.

A lot of different computing methods have been developed allowing coping with cumbersome quantitative methods that have to be used to determine the output states of the electromagnetic wave amplitude and polarization. The most successful and obvious is the Muller formalism, where matrix algebra is used to describe the amplitude and polarization transformations [38–41]. The optical elements are represented in the form of a T_{ij} matrices 4×4 that describe the polarizing elements, delay elements, and rotation matrix [39]. All values in this approach are real numbers.

The connection between light intensity Φ, the degree of polarization (p), and the polarization ellipse parameters (ψ, χ) (insert in the center of **Figure 3**) with the Stokes light parameters (vector **s**) is described with the algebraic expression:

$$\mathbf{s} = \begin{bmatrix} s_0 \\ s_1 \\ s_2 \\ s_3 \end{bmatrix} = \begin{bmatrix} \Phi \\ \Phi\, p \cos 2\psi \cos 2\chi \\ \Phi\, p \sin 2\psi \cos 2\chi \\ \Phi\, p \sin 2\chi \end{bmatrix}. \tag{9}$$

In case where polarizers are arranged perpendicularly to the incident light rays and the rays are parallel to the Z-axis, the polarizer and the analyzer are described by the linear transformation according to the Muller formalism [39]:

$$\mathbf{P}(\theta) = \frac{1}{2} \begin{bmatrix} 1 & \cos 2\theta & \sin 2\theta & 0 \\ \cos 2\theta & \cos^2 2\theta & \cos 2\theta \sin 2\theta & 0 \\ \sin 2\theta & \cos 2\theta \sin 2\theta & \sin^2 2\theta & 0 \\ 0 & 0 & 0 & 0 \end{bmatrix}, \tag{10}$$

where θ is the angle of the fast axis of the polarizer measured from the X-axis to the Y-axis (**Figure 3**).

The quarter-wave plate and the photoelastic element are described by a matrix for linear delay [38]:

$$\mathbf{R}(\theta,\delta)=\begin{bmatrix}1 & 0 & 0 & 0\\ 0 & \cos^2 2\theta+\cos\delta\sin^2 2\theta & (1-\cos\delta)\cos 2\theta\sin 2\theta & -\cos\delta\sin 2\theta\\ 0 & (1-\cos\delta)\cos 2\theta\sin 2\theta & \cos\delta\cos^2 2\theta+\sin^2 2\theta & \cos 2\theta\sin\delta\\ 0 & \sin\delta\sin 2\theta & -\cos 2\theta\sin\delta & \cos\delta\end{bmatrix}. \quad (11)$$

The elements are arranged perpendicularly to the incident light rays, and the rays are parallel to the Z-axis. θ is the angle of the fast axis measured from axis X to axis Y (**Figure 3**). $\delta=\delta_y-\delta_x$ is the phase difference between the fast and the slow axes (the delay).

The Mueller matrix for refraction [41] in a PE is:

$$\mathbf{T}(\phi,\psi)=\frac{\sin 2\phi\sin 2\psi}{2(\sin\phi_+\cos\phi_-)^2}\begin{bmatrix}\cos^2\phi_-+1 & \cos^2\phi_--1 & 0 & 0\\ \cos^2\phi_--1 & \cos^2\phi_-+1 & 0 & 0\\ 0 & 0 & 2\cos\phi_- & 0\\ 0 & 0 & 0 & 2\cos\phi_-\end{bmatrix}. \quad (12)$$

where ϕ – incidence angle, ψ – refraction angle, $\phi_\pm=\phi\pm\psi$.

These matrices make it easy to study the dependence of the light intensity and polarization on the angles of all transducer elements optical axes. The results obtained make it possible to determine the tolerances for the inaccuracy of the optical elements installation. Here, for simplicity, the light rays were considered as plane wave rays that fall at right angles to the surface of each element of the optical layout. We neglected the point source of light. After substituting all Muller matrixes and taking the first components, the photocurrents I_1 and I_2 of the photodetectors PD1 and PD2 (**Figure 3**) take the form:

$$\begin{aligned}I_1&=q\frac{(4n)^2}{(n+1)^4}\Phi_0\langle\mathbf{e}_1,\mathbf{L}_1\mathbf{e}_1\rangle=4n^2q\Phi_0\frac{1+\sin\Delta_{PE}}{(n+1)^4},\\ I_2&=q\frac{(4n)^2}{(n+1)^4}\Phi_0\langle\mathbf{e}_1,\mathbf{L}_2\mathbf{e}_1\rangle=4n^2q\Phi_0\frac{1-\sin\Delta_{PE}}{(n+1)^4},\end{aligned} \quad (13)$$

where q – photodetector quantum efficiency, $n=n_2/n_1$ –relative refractive index, Φ_0 – light intensity, Δ_{PE} – phase shift caused by the light ray passed through the photoelastic element, $\mathbf{e}_1$ is the identity matrix, $\mathbf{L}_1$ and $\mathbf{L}_2$ are linear transformations:

$$\begin{aligned}\mathbf{L}_1&=\mathbf{P}\left(-\frac{\pi}{4}\right)\mathbf{R}(0,\Delta_{PE})\mathbf{R}\left(-\frac{\pi}{4},\frac{\pi}{2}\right)\mathbf{P}(0),\\ \mathbf{L}_2&=\mathbf{P}\left(\frac{\pi}{4}\right)\mathbf{R}(0,\Delta_{PE})\mathbf{R}\left(-\frac{\pi}{4},\frac{\pi}{2}\right)\mathbf{P}(0).\end{aligned} \quad (14)$$

4.2 Transducer output signal

The output signal I_{out} of the electronic circuit is

$$I_{out}=k\frac{I_1-I_2}{I_1+I_2}, \quad (15)$$

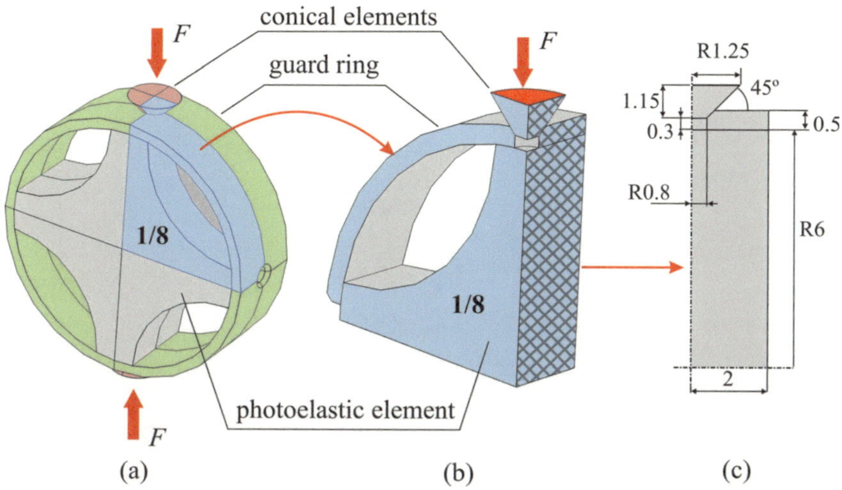

Figure 5.
(a) 3D model of the photoelastic element in the guard ring made of steel. Force F is applied to the top and bottom of the conical elements; (b) 1/8 part of the model; (c) model dimensions are indicated in millimeters [30].

where k is a proportionality factor determined by the transducer electrical circuit parameters. As a result, the signal, after the electronic circuit [36], takes the following form:

$$I_{out} = k \sin \Delta_{PE} \approx k\Delta_{PE}, \text{at } \Delta_{PE} \ll 1 \tag{16}$$

The equation shows that the change of output signal dI_{out} is directly proportional to the phase shift change $d\Delta_{PE}$ that is caused by the change of stresses in the photoelastic element.

4.3 Simulations of stresses and deformations in the photoelastic element

We used the COMSOL Multiphysics™ package and the finite-element method for the simulations of the spatial distribution of $\Delta\sigma = \sigma_y - \sigma_x$ in the PE, and PE deformation under the force F.

The 3D model of the PE encased in the steel guard ring is shown in **Figure 5a** [30, 31]. The upper and lower steel conical elements transmitted the external force to the guard ring. The embedded in COMSOL Multiphysics parameters of the material needed for the calculation of the deformation were used. Due to the PE symmetry, the simulation was only for the 1/8 part of it as shown in **Figure 5b**. The dimensions of this model part are shown in detail in **Figure 5c**.

5. Simulation and experimental results and analysis

The results of accurate numerical simulations and experiments are also detailed in [30, 31], here is a summary of them.

The strain gauge sensitivity determination was based on the experimental determination of the minimum detectable force and numerical simulation of the deformation occurring under the force action. We used the device "rhombus" with frictionless hinges to apply a calibrated force to the photoelastic element (**Figure 6a**).

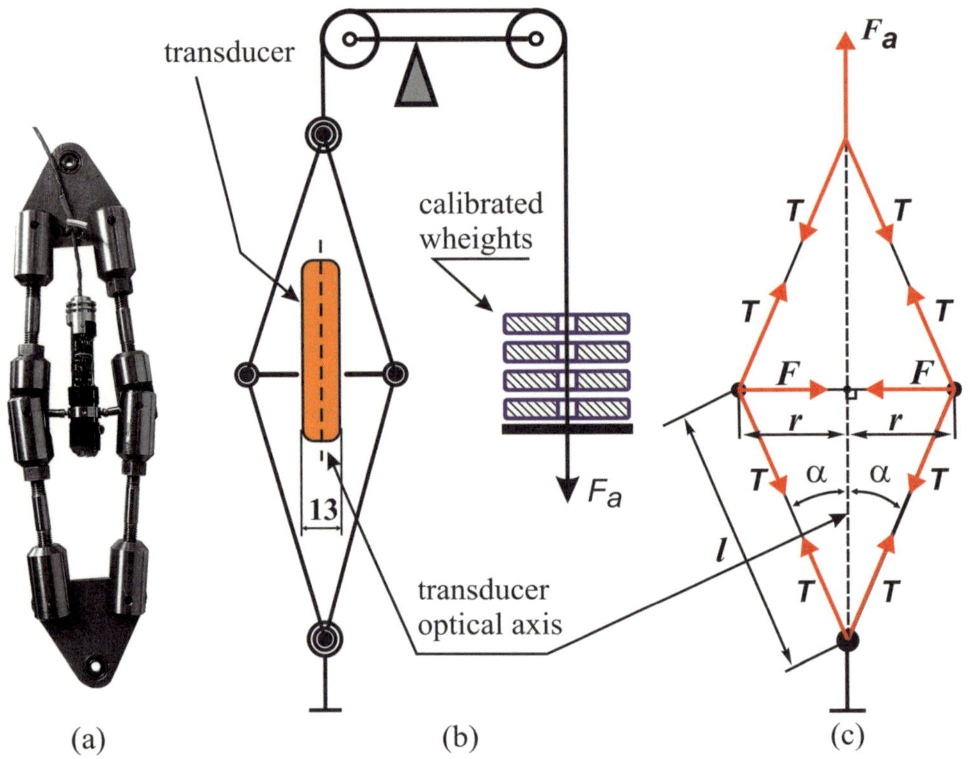

Figure 6.
(a) Rhombus photo with the transducer; (b) Force application scheme; (c) Diagram of forces in the rhombus with a fixed transducer upon the application of calibrated force F_a.

The rhombus with a fixed sensor was placed into the device for applying a calibrated force to the rhombus (**Figure 6b**). The rhombus was firmly restrained from one side and a calibrated force F_a was applied to the other side along the main axis of the rhombus symmetry pattern. The force was created by a lever mechanism with calibrated weights. The lever mechanism has the force transmission ratio of 1:8. The calibrated weight was 0.1 kgf. Thus, the weight applied to the rhombus was 0.8 kgf (7.848 N). The rhombus design ensured the force was applied to the PE in the direction perpendicular to the optical axis of the piezo-optical transducer (**Figure 6b**). To this end, the force from the rhombus was transmitting on the PE by means of conical tips that rested against the tapered holes in the guard ring (**Figure 5**). This joint provides weak stress distribution dependence in the photoelastic element on the deviation of the optical axis from the rhombus axis, due to the mobility of these elements relative to each other. In the experiments, this deviation did not exceed 1 degree, which gives the stress magnitude deviation in the PE working area (**Figure 3**) obtained by the numerical simulation is less than 0.02% and can be neglected.

A diagram of the forces generated in the rhombus with a fixed sensor upon calibrated force F_a application is shown in **Figure 6c**. T – tension force of the rhombus shoulder, l – length of the rhombus shoulder, r – radius of the photoelastic element in the guard ring, α – the angle between the shoulder and the vertical axis of the rhombus, F – the sought force applied to the photoelastic element perpendicular to the optical axis of the piezo-optical transducer.

The equations for the static forces are:

$$\begin{aligned} F_a - 2T\cos\alpha &= 0, \\ F - 2T\sin\alpha &= 0. \end{aligned} \tag{17}$$

Evaluating F we get:

$$F = F_a \mathrm{tg}\alpha = F_a \frac{r}{\sqrt{l^2 - r^2}}. \tag{18}$$

Substituting the values used: F_a = 7.848 N, r = 6.5 mm, l = 75 mm, we get the magnitude of the force applied to the photoelastic element:

$$F = 0.68 \text{ N}. \tag{19}$$

The minimum detectable force $F_{min}$was measured using the described experimental setup. The value obtained was used to determine the minimum detectable PE deformation $\varepsilon_{min} = dL/L$. Further, the gauge factor was obtained by numerical simulation (GF_{sim}) and was compared with experimentally measured (GF_{exp}). The dynamic range DR and transducer sensitivity S to force and deformation were determined as well.

5.1 Minimum detectable force

We used an analogue loop interface with current 20 mA according to standard IEC 62056–21/DIN 66258. The electric current in the analogue loop is independent of the cable resistance (its length), load resistance, EMF inductive interference, and supply voltage. Therefore, such an interface is more preferable for information transfer with remote control. The circuit allowed us to simultaneously power the transducer and generate the output signal in range 4–20 milliamps using a two-wire cable. The multimeter Agilent 34461A was used to measure the output signal and record it to a computer using the multimeter software.

Typical time dependence of the sensor output current I_{out} under the rhombus loading by the calibrated weights is shown in **Figure** 7 [30, 31]. The sensor load was as follows. First, the initial preload was applied to remove the backlash. Then the rhombus was sequentially loaded with four equal calibrated weights, each providing the force of F = 0.68 N. The output signal was averaged for the four weights. The averaged output signal magnitude corresponding to force F = 0.68 N was around 300 microamperes.

The random noise ΔI_{out} of the output current I_{out} was analyzed to calculate the minimum sensor sensitivity to the applied force (noise equivalent force). We used the first 20 seconds of the time dependence, before the preload (insert in **Figure** 7) to calculate the standard output signal magnitude deviation according to the normal probability distribution for the random error (Gaussian distribution). The experimental data processing yields the standard deviation magnitude σ = 0.1278 μA, which is the commonly accepted measurement accuracy. As a result, the force measurement accuracy or the minimum detectable force is:

$$F_{min} = 0.1278 [\mu\text{A}] \times 0.68[\text{N}]/300[\mu\text{A}] \approx 0.00029 \text{ N} = 0.29 \text{ mN}. \tag{20}$$

Furthermore, the magnitude of the PE deformation under the force F_{min} was calculated.

5.2 Photoelastic element deformations

To simplifying the simulation, due to the symmetry of the model, the exerted force is estimated 1/4 of the experimentally applied that is 0.29 mN/4 = 0.073 mN. The accurate simulations of applied static force 0.073 mN to the model

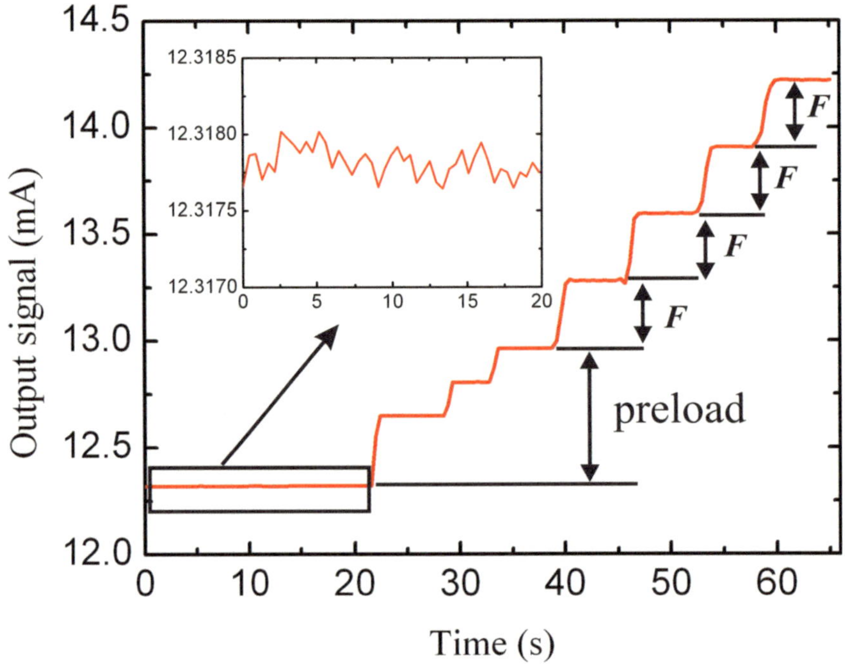

Figure 7.
Time dependence of the transducer output signal I_{out} when the load is applied consistently by means of identical calibrated weights.

(**Figure 5b**) yield the magnitude of the model deformation along the radius and along the axis of applied load $dL_{1/2} = -0.00175$ nm (i.e., this value by which the radius of the PE decreases along the applied force axis). To determine the PE diameter deformation, the result must be multiplied by 2. The resulting deformation is

$$dL = 0.0035 \text{ nm} = 3.5 \times 10^{-12} \text{ m}. \tag{21}$$

Thus, it is assumed that it is the minimum of the absolute deformation detectable by this transducer. And it corresponds to the relative deformation

$$\varepsilon_{\min} = dL/L = 3.5 \times 10^{-12} \text{ m}/12 \times 10^{-3} \text{ m} \approx 2.92 \times 10^{-10}, \tag{22}$$

where 12×10^{-3} m – the photoelastic element diameter.

This result is significantly better than that for the known industrial deformation sensors ($\varepsilon_{\min} > 10^{-8}$).

5.3 Piezo-optical transducer gauge factor

The accurate numerical simulation of the stresses which are rising in the PE working area under the applied force $F_{\min} = 0.29$ mN yields the magnitude:

$$\Delta\sigma = \sigma_y - \sigma_x = 17.11 \text{ Pa}. \tag{23}$$

The "effective" elasticity modulus E^* for present PE design can be calculated according to Hooke law:

$$E^* = \frac{\Delta\sigma}{\varepsilon} = \frac{17.11 \text{ Pa}}{2.92 \times 10^{-10}} = 58.6 \text{ GPa}. \tag{24}$$

The E^* value is somewhat smaller than the Young modulus value of fused quartz $E = 70$ GPa, due to the chosen PE design. Thus, Eq. (7) takes the form:

$$\mathrm{GF} = \frac{2\pi d}{\lambda} K E^*, \tag{25}$$

where the E^* value is determined by the PE design.

Now we need to define the gauge factor that works directly with photodetector output signals. If we take into account the output signal I_{out} with proportionality factor $k = 1$, we get

$$\frac{dI_{\text{out}}}{I_{\text{out}}} = \frac{I_1 - I_0}{I_0} = \frac{I_1 - I_2}{I_1 + I_2} = \mathrm{GF}_{\text{sim}} \frac{dL}{L}, \tag{26}$$

were I_0 is the I_1 signal from the photodetector PD1 without applied force ($\Delta_{\text{PE}} = 0$) and GF_{sim} is the simulated gauge factor value.

Taking into account the precise quarter-wave plate parameters (thickness $d_{\lambda/4}$ = 40 μm, $\Delta n = n_o - n_e$ =0.038 – characteristic of the quarter-wave plate birefringence), phase difference $\Delta = \Delta_{\text{PE}}$ and relative deformation magnitude $\varepsilon_{\min}$ =2.92 × 10^{-10}, the simulated piezo-optical transducer gauge factor GF_{sim} can be calculated:

$$\mathrm{GF}_{\text{sim}} = \frac{\sin\delta_{\lambda/4} \sin\Delta_{\text{PE}}}{\varepsilon_{\min}} = \frac{1}{\varepsilon_{\min}} \sin\left(\frac{2\pi d_{\lambda/4}}{\lambda}\Delta n\right) \sin\left(\frac{2\pi d}{\lambda} K \Delta\sigma\right) = 7389 \tag{27}$$

The simulated GF_{sim} value is somewhat smaller than the theoretical value $\mathrm{GF}_{\text{theor}}$ (Eq. (8)). This is due to the selected PE design, which determines the magnitude of the stresses $(\sigma_y - \sigma_x)$ in the PE working area for a given applied force value, and the $\sin\delta_{\lambda/4} = 0.947$ (not 1.0 for perfect quarter-wave plate).

5.4 Gauge factor dependence on photoelastic element shape

In order to determine the GF dependence on the PE shape (value of h in **Figure 4**), the relative deformation $\varepsilon = dL/L$ in Eq. (7) must be fixed in contrast to the method shown in **Figure 4**.

When varying the geometric parameters of the PE, the magnitude of the force was chosen so as to provide the same PE deformation in the direction of force application (see **Figure 4**), namely, dL = 100 nm. **Figure 8** shows the resulting dependence of GF on the parameter h [31]. It can be seen from **Figure 8** that the

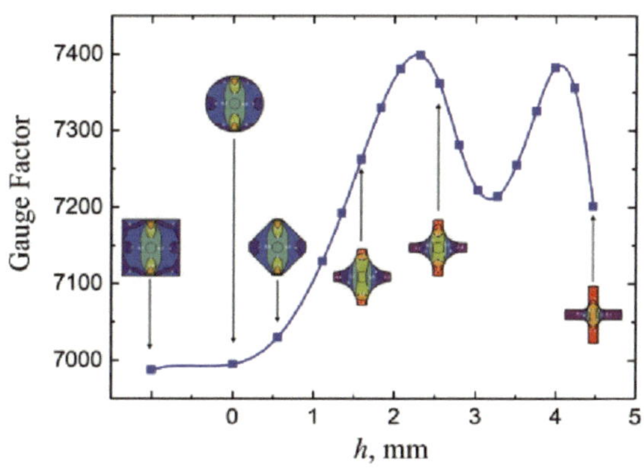

Figure 8.
Piezo-optical sensor gauge factor versus the parameter h.

dependence of GF on h is non-monotonic and contains two local maxima apparently due to the contribution of the nonlinearly changing shape of the PE side surfaces into its elastic properties. The changes in GF in the whole range of h were about 5.4% of the initial value, which is significantly less than the change in the stress difference $\Delta\sigma_{av}$ obtained in [32] and shown in **Figure 4**, which was almost 100%. This is due to the fact that as h increases, the PE stiffness (effective Young's modulus E^*) decreases in the direction of force application, which, in turn, leads to an increase in the relative deformation dL/L at the given force and a decrease in GF.

5.5 Piezo-optical transducer parameters

Experimental gauge factor. The direct measurement of the photocurrents (I_{PD1}, I_{PD2}) from the photodetectors (PD1, PD2 in **Figure 2**) yielded the experimental gauge factor GF_{exp} value

$$GF_{exp} = \frac{I_{PD1} - I_{PD2}}{I_{PD1} + I_{PD2}} = 7340. \tag{28}$$

This agrees well with the simulated gauge factor GF_{sim} and confirms the accuracy of the created transducer model.

Dynamic range. In our transducer design, as mentioned above, the PE has been affixed to the loading element in the initially stressed state that ensures the transducer operates at compressing and stretching deformation.

The transducer output signal varied from 4 to 20 mA. We set the initial output current value of 12 mA, corresponding to zero strain. The maximum change in the output signal I_{range} equal to ± 8 mA, then the dynamic range DR will be

$$DR = I_{range}/\sigma = 8 \times 10^{-3}\ A/0.1278 \times 10^{-6}\ A \approx 6.2 \times 10^4. \tag{29}$$

The resulting dynamic range value is much higher than the known values for strain gauges.

Sensitivity. The sensitivity S (the transfer function slope) was as follows:

$$S_F = 300\ \mu A/0.68\ N \approx 441\ \mu A/N \text{ to the force and} \tag{30}$$
$$S_L = 0.1278\ \mu A/0.0035\ nm \approx 36.6\ \mu A/nm \text{ to the deformation.}$$

6. Testing the piezo-optical transducer

For experimental verification of the claimed parameters of the piezo-optical sensor, we carried out comparative testing with the most sensitive of the strain-resistive gauge sensor Ultra Precision LowProfile™ Load Cell Interface Force™ (**Figure 9b**) [42] used only to calibrate the deadweight machines due to its complexity and high cost. The Load Cell, selected for comparison, had the nominal load of 2000 lbs. (907.185 kg) and is based on a strain-resistive sensor. Our Load Element was a parallelepiped made of hardened steel with a transducer installed in it as shown in **Figure 9a** and had a nominal load of 1000 kg, which is close the Load Cell nominal load. Tests took place in a certified laboratory Detroit Calibration Lab Trescal [43] laboratory using a deadweight machine of the National Institute of Standards and Technology (NIST) [44].

The Load Element and then Load Cell Interface Force™ were installed in a deadweight machine where they were consistently loaded by means of calibration weights. The most striking results of comparative tests were obtained under load

DOI: http://dx.doi.org/10.5772/intechopen.94082

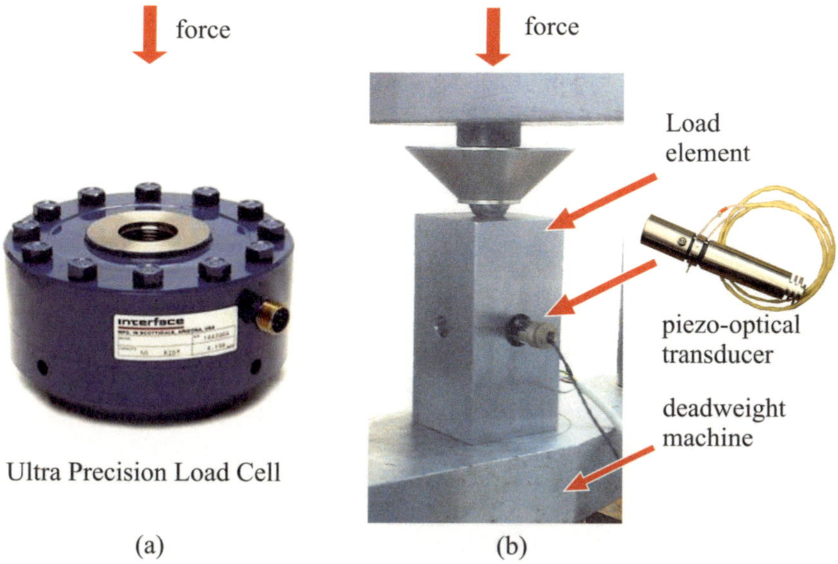

Figure 9.
(a) Ultra precision LowProfile™ Load Cell Interface force™; (b) Photos of our Load Element (left) with the piezo-optical transducer (right) and installed into a deadweight machine.

with the help of small weights, many times less than the nominal load value. The preload for both sensors was 110 lbs. and then the sensors were subsequently loaded with the calibrated weights from 1 gram to 100 grams The results are shown in **Figure 10**. The upper part of **Figure 10** corresponds to the presented piezo-optical transducer and the lower part – to the Load Cell Interface Force™. It can be seen that the piezo-optical transducer accuracy is approximately an order of magnitude higher than that for the Load Cell. This is less than the predicted calculations, and it

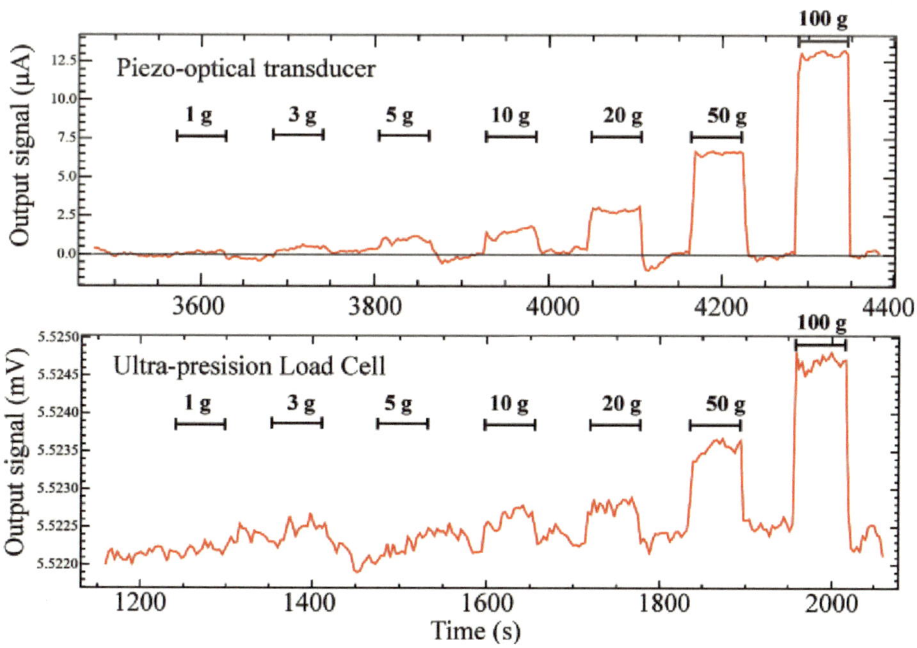

Figure 10.
Time dependence of the piezo-optical transducer (upper) and Load Cell (lower) output signals at the sequential load with calibrated weights 1, 3, 5, 10, 20, 50, 100 g.

is due to the fact that the Load Cell contains a vacuum chamber where a complex and expensive circuit is located to stabilize the output signal and reduce the noises. In our sensor, we used a design which was as simple as possible since the sensor is designed for a wide range of consumers. Nevertheless, this design showed higher sensitivity compared to the calibration Load Cell.

7. Discussion

The piezo-optical transducer operation was studied in detail theoretically, experimentally, as well as with the help of accurate numerical simulation. In order to compare the main parameters of sensors based on different physical principles, expressions for the gauge factors of strain-resistive, piezoelectric and fiber-optic sensors were proposed and analyzed. Despite the high piezoelectric modules of new piezoelectric materials (electroactive polymers), the piezoelectric sensor gauge factors are similar to strain-resistive sensor gauge factors.

It was shown that the piezo-optical sensor gauge factor, in contrast to sensors of other types, depends on the sensor design and can be improved by optimizing the PE design. The PE cruciform shape allows stresses to be concentrated in its small working volume because fused quartz used has no plastic deformation and the compressive damage threshold is very high. The piezo-optical quartz sensor gauge factor (7389), obtained by numerical simulation of stresses and deformations in the PE, is confirmed by the experimental results (7340) and is two to three orders of magnitude greater than the gauge factors of sensors based on other physical principles.

Table 2 shows that piezo-optical transducer is superior to the known industrially usable strain gauges. The high sensitivity of the piezo-optical sensor opens up new possibilities in problems of deformation measuring and stress analysis. For example, the use of only one such sensor makes it possible to control all parameters of the elevator movement: acceleration and deceleration, jerk, vibration, sound, according to International standard ISO 18738-1:2012 (E) Measurement of ride quality — Part 1: Lifts (elevators), as well as friction between the elevator car and the rails [45].

Parameter	Sensor type			
	Strain-resistive	Fiber-optic	Piezo-electric	Piezo-optical
Gauge Factor (GF)	2–4	0,78	0,1–36	> 7000
Dependence of GF on sensor design	no	no	no	yes
Deformation-to-current transfer function slope, μA/μm	—	—	80	1100–50,000
Sensitivity to the relative deformation	$\sim 10^{-6}$	$\sim 10^{-6}$	$\sim 10^{-6}$	$< 6 \times 10^{-10}$
Measurement error, %	0.05–0.1	0.25–1.0	—	0.01–0.03
Dynamic range	10^3–10^4	10^3	10^3	$> 5 \times 10^4$
Hysteresis, %	0.5	no	no	no
Overload, % of nominal	< 20	—	—	300–1000
Parameters degradation	yes	no	yes	no
Type of measured loads	Dynamic, static	Static	Dynamic	Dynamic, static

Table 2.
Comparison of some basic strain sensors parameters.

The sensor can be used in all cases where winch mechanisms are used, for example, in mines, escalators, moving walks, conveyors, cranes, etc.

Due to its high sensitivity, the sensor can be used for remote deformation monitoring by mounting at a certain distance away from the measured deformations zone: in bridge structures, cars and railway wagons weight remote control, liquids and gases flow control.

8. Conclusions

The theoretical, technological and design foundations for the highly sensitive piezo-optical transducers creation for strain gauges have been developed. It has been shown experimentally that such sensors have:

- absence of hysteresis within $\approx 1.7 \times 10^{-5}$ of the nominal load;
- high sensitivity to mechanical stresses, significantly exceeding the sensitivity of strain-resistive, piezoelectric and fiber-optic gauges and allowing to register the value of force less than 3×10^{-4} N, with a transfer function slope of $\approx$ 440 μA/N and $\approx$37 μA/nm;
- wide dynamic range, up to 6×10^4;
- resistance to overload;
- new functionalities corresponding to the sensitivity to relative deformations less than 10^{-9}; the specified sensitivity is documented by tests in certified calibration laboratory Trescal (Detroit, USA) and other testing laboratories.

Acknowledgements

The authors are grateful to Alex Zaguskin, AZ Enterprise L.L.C. for his help in organizing the transducer test in the Detroit Calibration Lab Trescal, as well as to Robert Bolthouse for his help with the tests.

Author details

Andrey G. Paulish[1,2,3*], Peter S. Zagubisalo[1], Sergey M. Churilov[1], Vladimir N. Barakov[4], Mikhail A. Pavlov[4] and Alexander V. Poyarkov[4]

1 Novosibirsk Branch of Rzhanov Institute of Semiconductor Physics, SBRAS, Technological Design Institute of Applied Microelectronics, Novosibirsk, Russian Federation

2 Novosibirsk State Technical University, Novosibirsk, Russian Federation

3 National Research Tomsk State University, Tomsk, Russian Federation

4 Firma PODIY, Ltd., Moscow, Russian Federation

*Address all correspondence to: paulish63@ngs.ru

References

[1] Vinter B, Rosencher E, Piva P. G. Optoelectronics. Cambridge University Press CUP; 2002. 744 p. ISBN: 10: 0521778131/ISBN: 13: 9780521778138.

[2] Parker M A. Physics of Optoelectronics. Boca Raton. CRC Press Taylor & Francis Group; 2005. 766 p. ISBN:13: 9780824-53856

[3] Paulish A, Fedorinin V, Levina A. Optoelectronics in Measurement of Physical Magnitudes. New York: Nova Science Publishers, Inc.; 2011. 63 p. ISBN: 9781617610943

[4] Kasap S. O. Optoelectronics & photonics: principles & practices. 2nd English ed. Boston: Pearson Education Limited; 2013. 544 p. ISBN:13: 9780132151498.

[5] Dutta N. K, Zhang X. Optoelectronic Devices. Singapore: World Scientific Publishing Company; 2018. 588 p. ISBN: 13: 9789813236691.

[6] Doyle J. Modern experimental stress analysis. Chichester: J.Wiley& Sons; 2004. p. 158–170. DOI: 10.1002/0470861584.ch3.

[7] Kobayashi A. Handbook on Experimental Mechanics. Englewood Cliffs, NJ: Prentice-Hall, Inc; 1987. 1002 p. ISBN:13: 9780133777062

[8] Qiu L, Deng X, Yuan S, Huang Y, Ren Y. Impact monitoring for aircraft smart composite skins based on a lightweight sensor network and characteristic digital sequences. IEEE Sensors Journal. 2018;**18**:2218-2244. DOI: 10.3390/s18072218

[9] de Inestrillas Á, Camarena F, Cabo M, Barreiro J, Reig A. Design and performance of a metal-shielded piezoelectric sensor. IEEE Sensors Journal. 2017;**17**:1284. DOI: 10.3390/s17061284

[10] Vitola J, Pozo F, Tibaduiza D and Anaya M. distributed piezoelectric sensor system for damage identification in structures subjected to temperature changes. IEEE Sensors Journal 2017;**17**: 1252. DOI: 10.3390/s17061252.

[11] Li M, Cheng W, Chen J, Xie R, Li X. A high performance piezoelectric sensor for dynamic force monitoring of landslide. IEEE Sensors Journal. 2017;**17**: 394. DOI: 10.3390/s17020394

[12] Rade D, de Albuquerque E, Figueira L, Carvalho J. Piezoelectric driving of vibration conveyors: An experimental assessment. IEEE Sensors Journal. 2013;**13**: 9174-9182. DOI: 10.3390/s130709174

[13] Wei P, Han X, Xia D, Liu T, Lang H. Novel fiber-optic ring acoustic emission sensor. Sensors. 2018;**1**:215. DOI: 10.3390/s18010215

[14] Liang H, Jia P, Liu J, Fang G, Li Z, Hong Y, et al. Diaphragm-free fiber-optic Fabry-Perot interferometric gas pressure sensor for high temperature application. IEEE Sensors Journal. 2018; **18**:1011. DOI: 10.3390/s18041011

[15] Sbarufatti C, Beligni A, Gilioli A, Ferrario M, Mattarei M, Martinelli M, et al. Strain wave acquisition by a fiber optic coherent sensor for impact monitoring. Materials. 2017;**10**:794-809. DOI: 10.3390/ma10070794

[16] Yang N, Su J, Fan Z, Qiu Q. High precision temperature insensitive strain sensor based on fiber-optic delay. IEEE Sensors Journal. 2017;**17**:1005. DOI: 10.3390/s17051005

[17] Peled Y, Motil A, Kressel I, Tur M. Monitoring the propagation of mechanical waves using an optic fiber distributed and dynamic strain sensor based on BOTDA. Optics Express. 2013; **21**:10697-10705. DOI: 10.1364/OE.21.010697

[18] Wild G, Hinckley S. Acousto-ultrasonic optical fiber sensors: Overview and state-of-the-art. IEEE Sensors Journal. 2008;**8**:1184-1193. DOI: 10.1109/JSEN.2008.926894

[19] Drissi-Habti M, Raman V, Khadour A, Timorian S. Fiber optic sensor embedment study for multi-parameter strain sensing. IEEE Sensors Journal. 2017;**17**:667. DOI: 10.3390/s17040667

[20] Frocht M. Photoelasticity. New York: Wiley-Interscience; 1948. Vol. 1. p. 432

[21] Piezooptical measuring transducer, by Slezinger I, Belitsky G, Shiryaev V, Mironov J. Patent US 4010632. 1977, Mar. 08. [Internet] Available: https://patents.google.com/patent/US4010632A/ja [Accessed: 2020-06-01].

[22] Photoelastic measuring transducer and accelerometer based thereon, by Trainer M. Patent US 4648274. 1987, Mar. 10. [Internet]. Available: https://patents.google.com/patent/US4010632A/ja [Accessed: 2020-06-01].

[23] Photoelastic neural torque sensor, by Discenzo F. Patent US 5723794. 1998, Mar. 03. [Internet]. Available:https://www.google.com/patents/US5723794 [Accessed: 2020-06-01].

[24] System and method for sensing torque on a rotating shaft, by Discenzo F. Patent US 6948381. 2005, Sep. 27. [Internet] Available: http://www.google.ch/patents/US6948381 [Accessed: 2020-06-01].

[25] Klerex T, Gunter B. Deformation measurement: Fiber optic sensors from HBM. ELECTRONICS: Science, Technology, Business. 2008;**1**:76-78

[26] Gautschi G. Piezoelectric Sensorics: Force Strain Pressure Acceleration and Acoustic Emission Sensors Materials and Amplifiers. Berlin: Springer-Verlag; 2002. 264 p. ISBN: 13:9783662047323

[27] Lushcheikin G. New polymer-containing piezoelectric materials. Fizika Tverdogo Tela. 2006;**48**:963-964

[28] Coker G, Filon L. A Treatise on Photo-Elasticity. Cambridge: University Press; 1931. 624 p

[29] Richard P. Birefringence dispersion in fused silica for DUV lithography. Proceedings SPIE.Optical Microlithography XIV. 2001;**4346**. DOI: 10.1117/12.435666

[30] Paulish A, Zagubisalo P, Barakov V, Pavlov M, PoyarkovA. Piezo-optical transducer for high sensitive strain gauges. IEEE Sensors Journal 2018; **18**: 8318–8328. DOI: 10.1109/JSEN.2018.2865917.

[31] Paulish A, Zagubisalo P, Barakov V, Pavlov M. Experimental investigation of a Piezo-optical transducer for highly sensitive strain gauges. Optoelectronics, Instrumentation and Data Processing. 2018;**54**:175-180. DOI: 10.3103/S8756699018020097

[32] Paulish A, Zagubisalo P. A Photoelastic element for Piezooptic strain gauges. Technical Physics Letters. 2015;**41**:632-634. DOI: 10.1134/S1063785015070135

[33] Musgraves JD, Hu J, Calvez L, editors. Springer Handbook of Glass. Cham: Springer Nature Switzerland AG; 2019. 1841 p. ISBN: 9783319937267

[34] Strain transducer, by Paulish A, Barakov V, Poyarkov A. Patent RU 2564691.2014, Feb. 04. [Internet] Available from: https://new.fips.ru/archive/pat/2015full/2015.10.10/doc/runwc2/000/000/002/564/691/document.pdfhttps://new.fips.ru/archive/pat/2015full/2015.10.10/doc/runwc2/000/000/002/564/691/document.pdf [Accessed: 2020-06-01].

[35] Tensometric transducer, by Paulish A, Poyarkov A, Shatov V,

Rafailovich A, Fedorinin V. Patent US 8,887,577. 2010, Apr.23. [Internet] Available from: https://patentimages.storage.googleapis.com/bd/b9/d8/0c7298424d783d/US8887577.pdf [Accessed: 2020-06-01].

[36] Piezooptic converter signal processing device, by Paulish A, Pavlov M, Barakov V, Poyarkov A. Patent RU 2565856.2014, July 08. [Internet] Available: https://new.fips.ru/archive/pat/2015full/2015.10.20/doc/runwc1/000/000/002/565/856/document.pdf [Accessed: 2020-06-01].

[37] Driggers R G. Encyclopedia of Optical Engineering. Vol. 1. New York: Marcel Dekker Inc; 2003. p. 205

[38] Udd E. Fiber Optic Sensors: An Introduction for Engineers and Scientists. 2nd ed. Inc.: John Wiley & Sons; 2011. 512 p. ISBN: 9780470126844

[39] Theocaris P, Gdoutos E. Matrix Theory of Photoelasticity. New-York: Springer-Verlag; 1979. 352 p. ISBN: 0387088997

[40] Nye J. Physical Properties of Crystals, their Representation by Tensor and Matrix. Oxford: Clarendon Press; 1985. 329 p. ISBN: 019851165 5 (pbk.)

[41] Goldstein D. Polarized Light. 2nd ed. New York: Marcel Dekker Inc.; 2003. 680 p. ISBN: 082474053-X

[42] Ultra Precision Low Profile ™ Load Cell Interface Force™. [Internet] Available: http://www.interfaceforce.com/index.php?mod=product&show=3 [Accessed: 2020-06-01].

[43] Detroit Calibration Lab Trescal. [Internet] Available: http://www.trescal.us/calibration-lab/michigan/detroit/detroit-calibration-lab [Accessed: 2020-06-01].

[44] Bartel T, Yaniv S, Seifarth R. Force measurement services at NIST: Equipment, procedures, and uncertainty. In: NCSL Workshop & Symposium. 1997. pp. 421-431

[45] Method for Lifting Device Motion Parameters Controlling, by Poyarkov A, Barakov V, Paulish A, Shatov V, Fedorinin V. Patent RU 2618862. 2015, October 15. [Internet] Available: https://www1.fips.ru/ofpstorage/Doc/IZPM/RUNWC1/000/000/002/618/862/%D0%98%D0%97-02618862-00001/document.pdf [Accessed: 2020-06-01].

Chapter 4

Fiber Optic Vibration Sensors

Putha Kishore, Dantala Dinakar and Manchineellu Padmavathi

Abstract

The sensors presented in this chapter are fiber optic intensity modulated vibrations sensors which are non-contact (extrinsic sensor) to the vibrating object. Three sensors presented make use of non-contact vibration measurement method with plastic fiber using distinct designs, improvement of the sensor response and advantages of one sensor over the other for diverse applications. First discussed about dual plastic optical fiber vibration sensor design and its response. Secondly, discussed about 1x2 fused coupler plastic optical fiber vibration sensor design with advantages over the first one. Finally, discussed about the 2x2 fused coupler plastic optical fiber vibration sensor design along with advantages than other two methods. At the end reported the final results with comparison.

Keywords: optical fiber, vibration sensor, intensity modulation, resolution, sensitivity and fiber Bragg grating

1. Introduction

It has been over five decades since the first emerged thought about the optical fibers could be used for sensing and measurement of the physical various parameters. Around 1960 the first patent was filed in the Photonic sensor, which is based on bifurcated bundle of fibers with half of the bundle used as transmitting fibers to illuminate on a reflecting surface and the other half of the bundle used as receiver to receive the reflected light from the reflector. The relative distance between the fiber bundles tip to the reflection precisely estimated by the suitable calibration process. In non-contact vibration sensing the Photonic sensors i.e. fiber optics have been continue for their unmatched offering of the results [1]. Fiber Optic sensors (FOS) provide many advantages over conventional sensors [2, 3], some of them as listed in **Table 1**.

In general, Fiber optics sensors are classified in to two groups: Intrinsic and Extrinsic sensors. In first type, the physical properties of the optical fiber itself can be used to convert effect of an environmental parameter on the optical fiber into a modulation of light parameters by passing through it. The light modulation parameters may be one of the following phase, intensity or polarization. Whereas, intrinsic FOS takes place within the optical fiber itself. Virtually, an environmental effect will be converted into light signal to be perturbed. In contrast, in extrinsic FOS, the optical fiber strictly used for carrying the information only that can be act as a black box to embed the information on an optical light, which is propagating through an optical fiber to a remote receiver. This black box usually contains optical elements such as a gas/liquid cells, a mechanical arm or so many other mechanisms that may

Conventional sensors	Fiber optic sensors
Bulk in size and weight dependent	Compact Size and light in weight
Most of them employ physically contact	Both contact and non-contact
Conductive	Nonconductive (Insulator)
Corrosive materials	Non-Corrosive materials
Chemically and electrically active	Chemically and electrically inert
Effected by Electromagnetic interference (EMI) and radio frequency interface (RFI)	Immune to EMI and RFI
Less bandwidth, no geometrical versatility and cost effective	Wide bandwidth, geometric versatility and economy
Low accuracy and sensitivity	Comparatively high accuracy and sensitivity
Difficult for distributed and multiplexed sensing	Can be easily used for distributed and multiplexed sensing
Not suitable for harsh environments and remote sensing applications	Suitable for harsh environment and remote monitoring systems can be made easily. Potentially easy to install

Table 1.
Comparison between conventional and fiber optic sensors.

cause modulation or transforming a light beam. Further, FOS sensors can also be classified based on their principle of working such as wavelength coding, Interferometric and Intensity modulated sensors. Intensity modulated FOS sensors are worked based intensity of light modulation with respect to the external perturbation. Phase modulated FOS sensors are passive in nature with optical elements that use phase change of the light field by the external perturbations, it is also called interferometric based sensor. The disadvantages of the optical fiber vibration sensors are the narrow frequency range of measurement and unfamiliarity to the end user. Thus, the fiber optic vibration sensors has required further research and development [4, 5].

1.1 Interferometric based vibration sensors

There exists few types of fiber optic interferometric vibration sensors such as Fabry-Perot, Mach-Zahnder, Michelson, and Sagnac [31] to interrogate the phase shift caused by vibration. In these sensors, the optical fiber as all-fiber interferometer which is usually a single mode optical fiber (SMF) rather than the multimode optical fiber (MMF). Because, the transfer function of SMF interferometers nearly reflects that of conventional interferometers. Whereas the transfer function of a MMF interferometer is independent of time owing to the more number of modes of the optical light in the optical fiber. Usually, phase variation in the interferometer can be produced either by an extrinsic or intrinsic effect. This phase can be encoded by the transfer function of the interferometer into modulation of light intensity at the photo-detector in a nonlinear method, using the usual interface cosine function. For most of the interferometers practical applications, a small sensor heads having a fiber optic Fabry-Perot (FP) interferometer along with a small length optical cavity are especially attractive. Because, they have the advantages of being simple in design, compact size, cheap, with lower cross sensitivity to ambient temperature and offers both high resolution and down lead insensitivity without the fading of polarization, usually faced by all in fiber optic interferometers [6]. A system using alternative EFPI arrangement is reported for the sensing of vibration, and its sensor head is shown in **Figure 1**. The sensor head uses a simple reflective configuration with an extrinsic FP cavity. An adjacent dual step-index MMF couple light into it

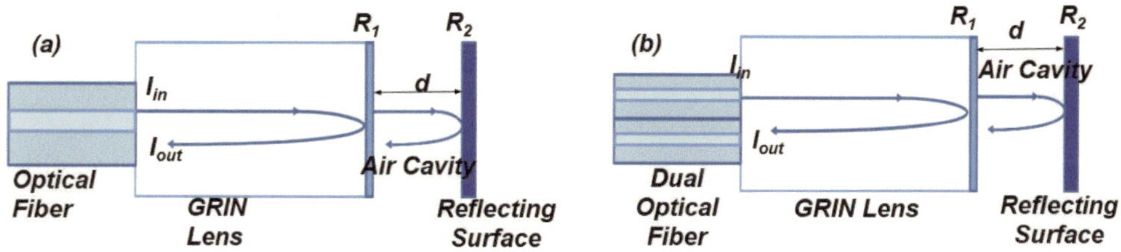

Figure 1.
The EFPI vibration sensor head of schematic (a) single fiber (b) dual fiber.

and out of the optical cavity. The light is incident by a low power LASER diode as a source. Here, a movable reflecting surface used as a transducer and a suitable pitch gradient index cylindrical (GRIN) lens has been used for efficient light guiding device between the input and output optical fibers. A partial reflecting coating (R_1) on the output face of the GRIN lens act as an interference reference. A high reflective surface (R_2) moves in sympathy with respect to the target object vibrations, it provides the interference signal. The FP cavity has a length of d in air as shown in **Figure 1**. These interferometric methods offer better performance but shows low stability, expensive, critical alignment, mechanical requirement because of their, need complex analysis (fringe counting) and are not well suitable for sensing the vibration at various testing points. These sensors require an electronically driven element to change the interferometer conditions. As a consequence, these sensors have a limited practical use. Thus, most of the recent optical fiber sensors are employing using intensity modulation only [7, 8].

1.2 Intensity modulated vibration sensors

Intensity modulated fiber optic sensor techniques have been studied and implemented for the last three decades. A wide range of fiber optic configurations are reported, like fiber optic microbending, reflected light coupling to optical fiber, direct fiber-to-fiber coupling, fiber Bragg gratings, and modified cladding of optical fiber. All these sensors are classified into two fundamental methods either in physically in-contact or non-contact with the vibrating target or not. Usually non-contact configurations use a reflecting signal for detecting the displacement or vibration while the other configurations use the trans-missive configuration, i.e., microbending. As a general rule, in the intensity modulated sensor configuration, the intensity of light from the source is modulated by the optical fiber; then it is guided through the optic fiber to the photo-detector, then the light intensity pulses are translated into equivalent electronic signal, and adequately processed. In most of the case, a referencing mechanism is required in order to eliminate the other noises and maintain the stable sensor calibration. Without using referencing signal, the fluctuations owing of the power in light source, noise induced by the connectors, couplers, or any other optical components in the sensing system can become the significant relative errors. In this section, some of the intensity modulated sensors are discussed. For few decades, so many fiber optic sensors (FOS) works based on intensity modulation techniques have been demonstrated [4, 5].

1.3 Microbending vibration sensor

The microbend optical fiber sensor is one of the earliest sensor reported which is works on the principle of the Intensity modulated. The sensing principle is works of the transmitted light power variation of as a function of applied physical variable like pressure/stress [38]. Generally, in this configuration, the amplitude of light intensity reduced by the cause of loss by the strain induced micro curvatures. The structure of the microbend fiber optic vibration sensor is shown in **Figure 2**. The sensing element (optical fiber) is sandwiched between a pair of strain induced plates having micro structure of saw tooth, it is capable to bend the optical fiber structure in a regular geometrical pattern with a periodicity of Λ. This deforms the fiber, with respect to an appropriate physical change (ΔE), owing to applied force (ΔF) to bent fiber, which cause the change in amplitude of the fiber (X) to vary its quantity (ΔX). The transmission coefficient of light which is propagating through the bent optical fiber (T_{p}) is changed by a quantity (ΔT_{p}), therefore

$$\Delta T = \left(\frac{\Delta T_p}{\Delta X}\right) D \Delta E \quad (1)$$

Here, D is a constant which is dependents on the physical change ΔE.

The deformation triggers a coupling of the light power from the optical fiber core guiding modes to higher order radiation (cladding) modes; which are easily perturbed by the surrounding medium. Both MMF and SMF have been used for the development of these sensors. In SMF microbend sensors, the maximum sensitivity is observed when the spatial bend frequency equals to the difference between the propagation constants of the fundamental mode and a discrete cladding mode [5, 9]. The microbending sensor has to be placed in-between the deformer plates to detect applied pressure. Denis Donlagic and Miha Zavrsnik reported a novel structural method by single-mode leads and multimode fiber (SMS) based on microbending on the multimode section of the optical fiber is shown in **Figure 3**. It exhibits higher sensitivity than classical microbend sensors [10].

In addressing the fiber strength issues, it is to be remembered that the deformer plates clamp the optical fiber. Therefore, a large stress can be produced on to the fiber. Suppose the deformer plates are brought very close together, the optical fiber may leads to the breakage. An empirical design instruction is to maintain the ratio of maximum applied stress to fiber break stress less than one to four. Since, microbend saw teeth push into the buffer coating of the optical, it should be very important to know the interaction of the buffer coating material and optical fiber with respect to the various testing properties Therefore, the principal disadvantage of the microbending sensors is that low accuracy.

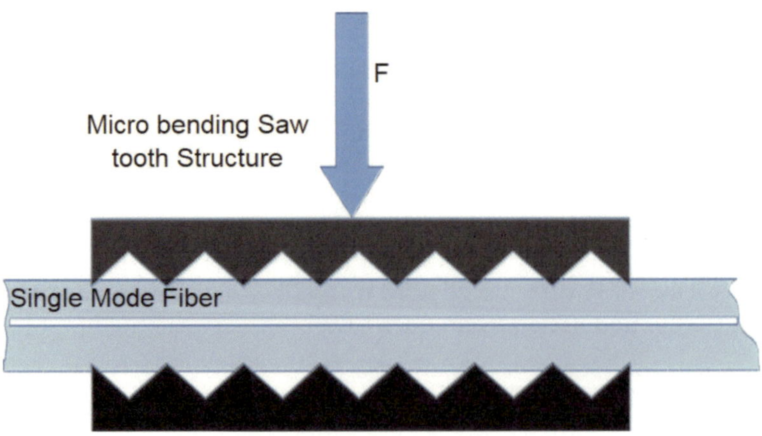

Figure 2.
Sensing principle of the microbend vibration sensor.

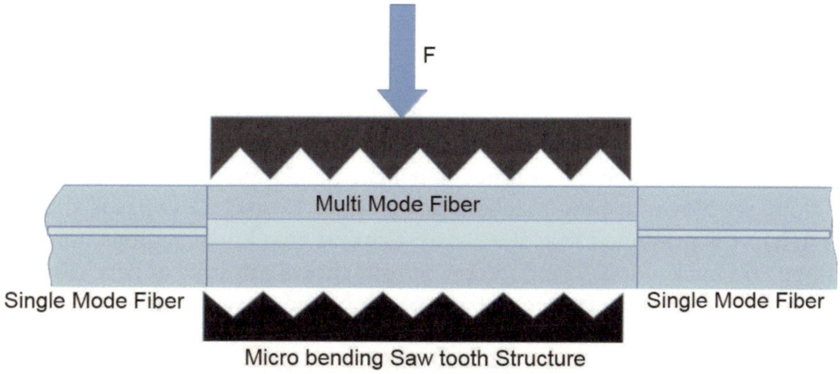

Figure 3.
Sensing principle of the SMS structure microbending sensor.

1.4 Non-contact vibration sensors

Most of the non-contact dynamic displacement sensors commonly can be used for the sensing of vibration. Here, a reflective mechanism is used for detecting the vibrations, where one optical fiber is used as a transmitting of the light source and another fiber is are used as a collector. The reflection of the surfaces of the target object can be minimized with help of data treatment methods. Binu et al., presented a simple, rugged, and cheapest non-contact intensity modulation based fiber optic sensor with configuration of two PMMA optical fibers cemented together [11]. The same design was proposed by the Yasin et al., [12]. The important benefits of this design is the low fabrication cost of the device. Although intensity modulated based fiber optic sensors are cheap and easy to fabricate, a weighty error in the measurement can be presented owing to effect of light source power variations. Losses owing to physical design and reflective planes outside of the measuring structure often effect on the accuracy of the final measurement. Fortunately, source light intensity fluctuations can be easily eliminate with referencing port.

Recently, Perrone et al., reported a low cost and high resolution using plastic optical fibers (POF) based on the reflected intensity modulation using dual POF. It is capable to measure the vibrations of up to several KHz by using an intensity modulation technique with a simple data processing and compensate the reflectivity of the vibrating surface. The received optical signal is incident onto the photo-detector and processed for the conversion. However this process is not user friendly and poses critical analysis process just like interferometric sensor. Those intensity modulated based fiber optic sensors are usually very cheap, easy to build and versatile in structures [13].

Further, an intensity modulated based displacement sensor reported, which is working on guiding optical light through the optical fiber onto a reflecting surface. Lewis et al., demonstrated the configuration in which the reflected light is collected by the same incident optical fiber [14]. The transducer itself can be a simple reflecting surface which is attached to the surface of a vibrating object. This fiber optic vibration sensor is a low cost and reliable, which is alternative for non-contact vibration detection with high-resolution frequency analysis. However, the multimode fiber having low dimension is limiting the practical application of the sensor. Because the sensor was positioned perpendicular to the vibrating body, it is difficult to align and maintain the sensor position constant at this dimensions.

This chapter have a more concentration on the plastic optical fiber vibration sensors design and development for the last few decades.

2. Dual optical fibers

The sensing head consists of two fibers made up of PMMA (Polymethyle Methacrylate) [15, 16], where one fiber acts as a transmitting fiber (TF) and other fiber acts as a receiving fiber (RF) which are bundled together parallel to each other [17]. The schematic setup is show in **Figure 4(a)**, the displacement response of the sensor along with overlapping mechanism between TF and RF cones is shown in **Figure 4(b)**. It is predicted that the sensor exhibits two linear regions namely the front slope and back slope. The detector output shows minimal at zero distance (Z = 0) between reflecting target and sensor probe, because the reflecting light cone of the TF does not reach the receiving cone of RF. As the distance from the sensor probe increases (Z < Zmax), the cone size of the transmitted light on the reflecting surface also increases, thereby causing the overlap with the RF cone which leads to a negligible output voltage. Further an increase in distance leads to larger overlap leading to rise in the voltage. This response reaches a maximum voltage where complete overlapping of the RF with TF reflecting cone occur (Z = Zmax), and then the output starts decreasing even though the distance increased (Z > Zmax). Because, the size of the reflecting light cone increases to very large which leading to decrease in power density, whereas the overlapping area remains constant [18, 19].

The front slope exhibits high linearity in the range of about 350-800 μm with a sensitivity of 4.786 mv/μm. A dark region of 350 μm is observed in the characteristic curve, where the intensity of light is not linear with the displacement for a small distance due to the cladding of the optical fiber. On the other hand, the back slope shows high linearity in the range between 1600 μm and 2600 μm with sensitivity of 1.696 mv/μm. Therefore, the front slope exhibits relatively high sensitivity but over small measurable range compared to the back slope and is better suited for the measurement of amplitude of vibration in micro level [20, 21].

2.1 Theory

According to the light intensity distribution function, the irradiance of emitted light from transmitting fiber is expressed as [20, 21].

$$I(r,z) = \frac{2P(z)}{\pi\omega^2} exp\left[\frac{-2r^2}{\omega^2(z)}\right] \tag{2}$$

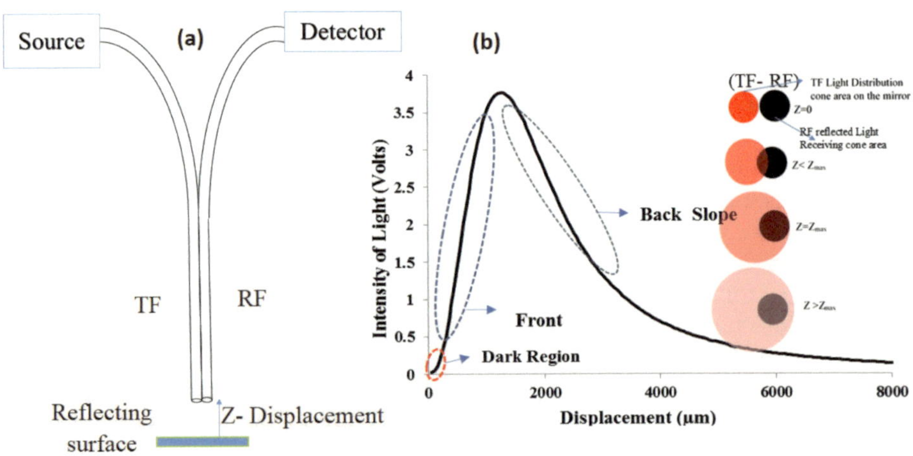

Figure 4.
(a) Schematic of the sensor and (b) displacement response of the sensor.

Where r and z represent the radial and longitudinal coordinates respectively, and $\omega(z)$ is the beam radius which can be expressed as a function of z given below

$$\omega(z) = \omega_o \sqrt{1 + \left(\frac{z}{z_R}\right)^2} \quad (3)$$

where Z_R is expressed as

$$Z_R = \sqrt{\frac{\pi \omega_o^2}{\lambda}} \quad (4)$$

where, Z_R is the Rayleigh range and ω_o is the beam waist radius. The quantity of reflected intensity of light power received by the RF from the target is solved by taking the integration of the irradiance over the core area S_r

$$\mathrm{P}(z) = \int_{sr}^{0} I(r,z) ds_r \quad (5)$$

The quantity of reflected intensity of light power collected by the RF is a function of the displacement between probe and target (reflecting surface) which can be expressed as [6, 7].

$$P(z) = \frac{2P_E}{\pi \omega^2(z)} \int_{y=-R_r}^{R_r} \int_{x=m_1}^{m_2} exp\left[\frac{-2(x^2+y^2)}{\omega^2(z)}\right] dxdy \quad (6)$$

Where $m_1 = R_t + R_r + R_d - \sqrt{R_r^2 - y^2}$ and $m_2 = R_t + R_r + R_d + \sqrt{R_r^2 - y^2}$, PE is the power of light from the TF incident on the reflector, R_t and R_r are the core radius of TF and RF respectively and Rd. is the distance between the centers of RF and TF cores.

A simple photo-detection circuit is used for the conversion of the light intensity into equivalent output voltage. Generally, the output voltage with respect to the intensity of light incident on photo-detector is given by [8, 9].

$$V_{out} = R_\lambda P R_E \quad (7)$$

Where $R_\lambda = \eta g \frac{\lambda}{1.24}$ is the photo-detector responsivity, R_E is the feedback resistance, η, λ and g are the quantum efficiency, wavelength of the incident light and photoconductive gain respectively. For a given photo detector the values of η (<1) and g are constant, therefore the responsivity (sensitivity) depends on the wavelength of the light source. Thus light source and photo-detector should be matched.

2.2 Experimental setup

The schematic of the experimental setup of the fiber optic vibration sensor is shown in **Figure 5**. The sensing head consists of two PMMA fibers, with constant diameters bundled together in parallel. A commercial speaker/PZT can be used as a vibrator to test the response of the FOS.

A thin plastic reflector of thickness100μm is glued at the center of the speaker to act as reflecting surface. An LED is used as a light source which should be matched to the optical transmission window of the PMMA fiber. The LED is housed in a special package which facilitates perfect holding and provide maximum coupling of

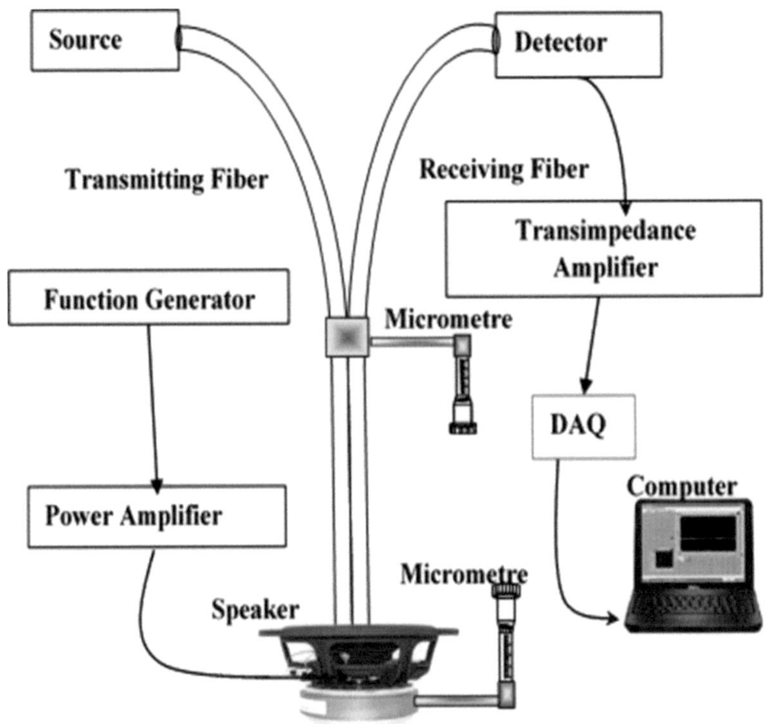

Figure 5.
Schematic experimental setup of the dual POF vibration sensor.

light into the fiber. One end of the transmitting fiber is fixed to the housing of the LED. The frequency and amplitude of vibrations of the speaker is controlled through a Function Generator and Power Amplifier.

The design of a dual plastic optical fiber (POF) vibration sensor using different fiber pair combinations reported along with necessary theory and experimental results. From displacement response of all the combinations, it is evident that the sensitivity of the sensor increases as the diameter of the fiber decreases and vice versa. The vibration response of the sensor for all the combinations reveals that when the fiber diameter of either TF or RF decreases, the frequency range is increased and resolution is improved. Further, the dynamic range, and the range of frequency can be optimized by using the suitable diameters of the fiber. Moreover, the dark region of the sensor can be minimized by choosing the diameter of the fiber as small as possible. The fiber combination of lower diameter shows better response than any combination and it exhibits the high frequency response with high resolution when compare with others. However, this dark region is one of the major drawback of the sensor configuration [20, 21].

3. Fiber optic fused 1x2 coupler

The sensor system consists of a 3 dB fiber optic 1x2 coupler made of PMMA fiber having three ports, in which the first port is used as a sensing probe while the second port is for coupling of light from the light source, and the third port is used to direct the reflected light, to be incident on the photo-detector [22–24]. The principle of the vibration measurement is based on intensity modulation with respect to the displacement between the reflecting surface glued on the vibrating

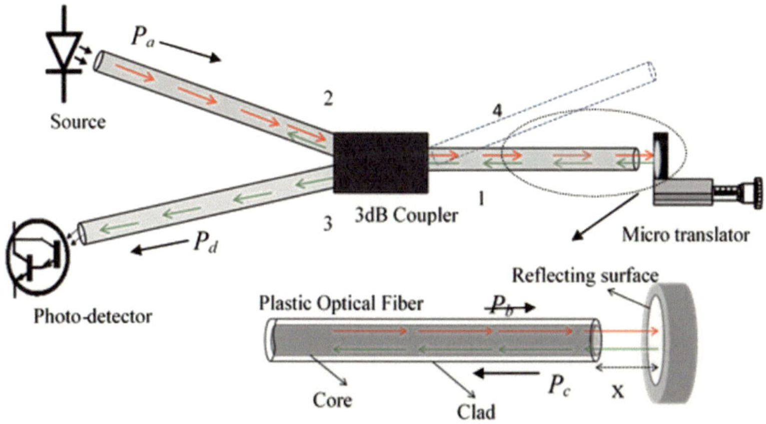

Figure 6.
Working principle of the fiber optic coupler vibration sensor.

target and the sensing fiber port [25–28]. The schematic diagram of the sensing principle is illustrated in **Figure 6**. Light from the sensing fiber is allowed to be incident on the reflecting surface (glued on the front surface of the micro translation stage) which is kept at a distance of x from tip of the sensing fiber (port1) and the reflected light is allowed to be coupled back into the same fiber.

3.1 Theory

If P_a, P_b, P_c and P_d.represent the power of light coupled to port 2, light incident on the reflector through port1, the light reflected from the reflector coupled back into port1 and the light power received by the photo-detector via port3 respectively, then, the light transmitted from the source through the fiber to the sensing fiber port1 can be given by [29, 30].

$$P_b = (1 - cr)\left(10^{-0.1L} - 10^{-0.1D}\right)P_a \tag{8}$$

Where cr, L and D are coupling ratio, excess loss and directivity of the optical fiber coupler respectively.

If the reflector is kept parallel to the sensing fiber cross-section, the power of light coupled back and received by the sensing fiber probe can be expressed as

$$P_c = P_i\left(1 - exp\left(-\frac{2a}{W^2(x)}\right)\right) \tag{9}$$

Where $P_i = kP_b$ is the light power coupled to the sensing fiber at $x = 0$, a is the core radius of the fiber, $W(x) = 2xtan(\theta) + a$, k = 1.15 and $\theta = \sin^{-1}NA$ is the divergence angle of the optical fiber [30].

Substituting Eq. (8) into (9), we have

$$P_c = k(1 - cr)\left(10^{-0.1L} - 10^{-0.1D}\right)P_a\left(1 - exp\left(-\frac{2a}{W^2(x)}\right)\right) \tag{10}$$

Finally, the light power detected by the photo-detector from the sensing port through the port3 can be written as

$$P_d = cr\left(10^{-0.1L} - 10^{-0.1D}\right)P_c \tag{11}$$

Substituting $W(x)$, Eq. (10) into (11) yield

$$P_d = k\,cr(1-cr)\left(10^{-0.1L} - 10^{-0.1D}\right)^2\left(1 - exp\left(-\frac{2}{\left(\frac{2xtan(\theta)}{a}+1\right)^2}\right)\right)P_a \qquad (12)$$

Therefore

$$P_d = P\left(1 - exp\left(-\frac{2}{\left(\frac{2xtan(\theta)}{a}+1\right)^2}\right)\right)P_a \qquad (13)$$

Where

$$P = k\,cr(1-cr)\left(10^{-0.1L} - 10^{-0.1D}\right)^2$$

For large value of $\frac{2xtan(\theta)}{a}$, Eq. (13) can be written as

$$P_d = \frac{P}{2}\left(\frac{(2a)^2}{(2xtan(\theta))^2}\right)P_a \qquad (14)$$

This equation is the correlation function of the displacement sensor with multimode fiber coupler. It states that the power received by the photo-detector is directly proportional to the square of the diameter of the fiber and is inversely proportional to the square of the distance between the sensor head and the reflecting surface.

3.2 Experiment

This simple sensor configuration can eliminates the presence of dark region and it exhibits only single slope that enables easy setup when compared to other configurations [31–33].

Figure 7 illustrates the schematic experimental setup of the fiber optic fused 1x2 coupler as a vibration sensor. It consists of a LED source of suitable wavelength, which is driven by a simple circuit with regulated power supply. A 3 dB Plastic

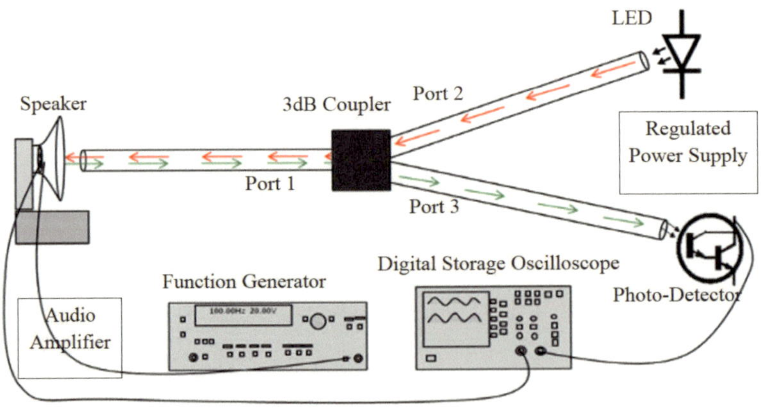

Figure 7.
The 1x2 fiber optic coupler vibration sensor experimental setup.

optical fiber 1x2-coupler is used to configure the sensor to detect the vibration. A photo-detector along with detection circuit is used to convert the light intensity into equivalent electrical signal. A synthesized function generator and a commercial speaker with a calibrated reflector attached at the center of it are used to test the sensor response for vibration. To record and monitor the vibration of the speaker at different frequencies and amplitudes, a digital storage oscilloscope is used. The whole experimental setup is installed on a vibration less table to eliminate the ground vibrations [31–33].

The calibration of the sensor for the measurement of amplitude of vibrations has been carried out. A weightless plastic reflector is glued on the surface of a rectangular block fixed to micro translation stage which is positioned perpendicular to the sensing head of the sensor. A digital Multimeter is used to measure the photo-detector output light in terms of voltage with respect to the displacement between the reflector and the sensing head (port1). **Figure 8** shows the experimental and theoretical displacement characteristic curves using Eq. (14). As shown in **Figure 8**, the linear region in the range of 0-1000 μm. The weightless reflector is now glued on to the speaker diaphragm and is placed perpendicular to the sensor probe (port1). The distance between the speaker and the sensor head is fixed within the linear region of the displacement curve shown in **Figure 8**. The light from the LED is coupled to the port2 of the coupler and is directed to the port1. The light incident on the reflector through port1 gets reflected back while modulated in response to the vibration, and is received by the same fiber (port1). The light power received by the photo-detector is then converted into its equivalent voltage signal by a simple receiving circuit and is recorded or stored by the DSO. FFT technique is implemented for the conversion of the time domain signal into frequency domain signal to analyze the vibration in terms of frequency and amplitude. The experiment is repeated for different frequencies and amplitude of vibrations to detect the maximum frequency and amplitude resolution that can be measured by the designed sensor and also to test the reliability of the sensor.

The experimental is setup on a vibration less table. The speaker is allowed to vibrate by a sine wave (CH1) through the Signal Generator and the response of the sensor (CH2) is recorded using DSO for different frequencies. The FFT of these signals gives the frequencies of the applied signal and output of the sensor. It is evident from the figure that, there is a perfect agreement between the applied signal and response of the sensor. The amplitude d_p of displacement can be computed from the knowledge of the peak to peak voltage of the output signal and the slope of

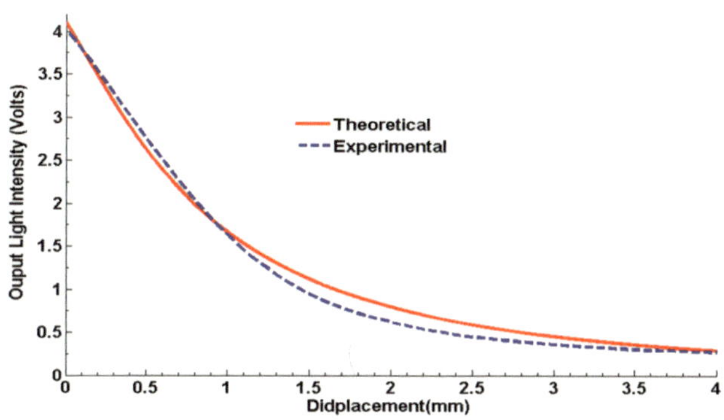

Figure 8.
The displacement characteristic response of the 1x2 fiber optic coupler as vibration sensor.

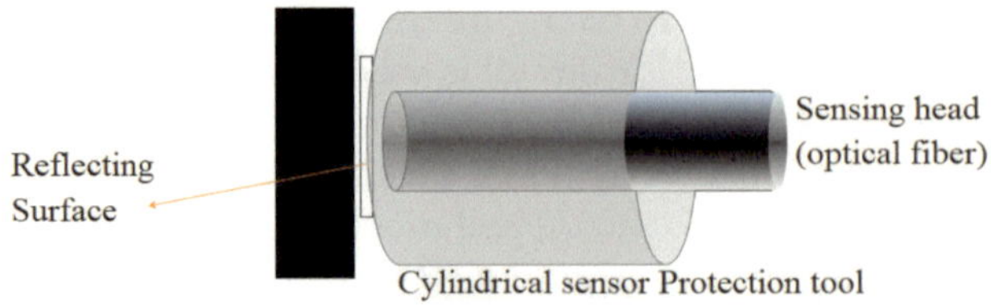

Figure 9.
Arrangement of the cylindrical sensor protection tool on the reflecting surface.

the calibration curve. For a given frequency f_p the peak velocity v_p and peak acceleration a_p of vibrating body can be computed by [31–33].

$$v_p = (2\pi) f_p d_p \tag{15}$$

$$a_p = (2\pi)^2 f_p^2 d_p \tag{16}$$

The possibilities of errors which might be present in measurement can be, light source fluctuations, stray light effect and dust formation on the mirrors. To reduce the fluctuations in the source of light a standard regulated power supply can be used. A hollow cylindrical protection tool is arranged surrounding the reflector to protect it from the stray light interference with the light from source and to reduce the dust formation on the mirror as shown in **Figure 9**. The sensor is positioned very close to the vibrating target within the linear sensing region and it does not require any special optics for enabling its use for sensing applications in embedded situations [31–33].

4. Fiber optic fused 2x2 coupler

In this second part discussed the design of the fiber optic 2x2 fused coupler as a vibration sensor. Study the displacement response and vibration response of the sensor. Implementation of the rational output method to improve response of the sensor than the 1x2 coupler [34, 35].

4.1 Displacement response of the sensor

Figure 10 illustrates the schematic of the proposed plastic multimode fiber optic 2x2 fused coupler made of Poly methyl methacrylate having a split ratio of 80:20 as a vibration sensor.

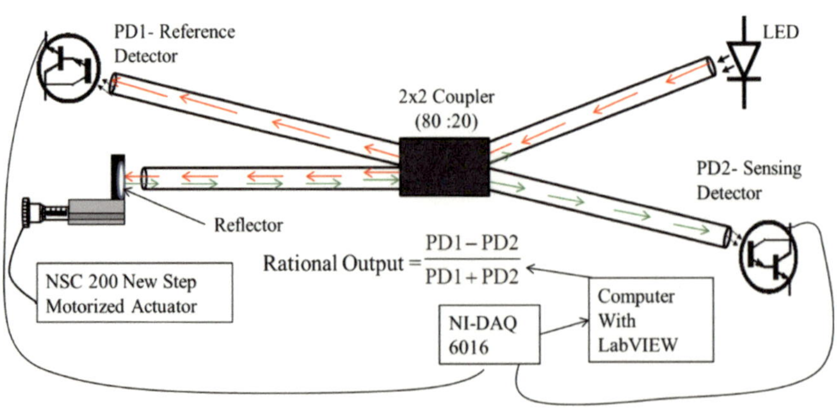

Figure 10.
Schematic of experimental setup for the displacement response of the sensor.

The sensor consists of an LED used as a light source and two numbers of photo-Darlington detectors (PD) of high sensitivity housed in a connector less package used to detect the intensity of light at reference and sensing ends. A simple photo detection circuit is developed for conversion of the modulated intensity of light into its equivalent output voltage signal and a DAQ is employed to record the time domain signals (TDS) corresponding to the reference and sensing arms from which the rational output (RO) is calculated.

The fiber optic fused 2x2 coupler has core/cladding diameters of 980/1000 μm, with split ratio of 80:20, and having four ports. All the ports of the coupler are used for vibration detection. The light from the LED and coupled to the port2 is split into the ratio of 80:20; and one part of light (80%) is transmitted through the port1, which act as a sensing probe and the other part (20%) is directed towards PD1 through port4 which is used as a reference signal [36, 37]. The light through port1 projects onto a weightless plastic reflector having the reflectivity of 40% which is attached to the center of the speaker diaphragm (the vibrating object). The reflected light modulations corresponds to vibration is recoupled into the same fiber (port1) and is directed to be incident on the PD2 through port3. In order to avoid the effect of power fluctuations in light source and bending losses in optical fiber, a reliable method of the rational output (RO) of PD1 and PD2 is taken into consideration and is expressed as [38–40].

$$\text{Rational Ouput (RO)} = \frac{\text{PD1} - \text{PD2}}{\text{PD1} + \text{PD2}} \tag{17}$$

A Step Motorized Actuator has been used to move the reflector attached to the micrometer stage to and fro from the sensing probe with a step size of 1 μm over a dynamic range of 4 mm. The experimental results depicts that the sensor displacement characteristic curve is presented in **Figure 11**, follows the inverse square law given by Eq. (14) and the linear region of this curve is used for the vibration measurement. It can be observed that the characteristic curve representing the response of PD2 with respect to the displacement of the reflector from the sensing probe (**Figure 11**) has a linear region of about 1 mm with a sensitivity of 2.1 mV/μm, whereas the response curve representing the RO has a sensitivity of 0.36a.u/mm. This linear region of both responses can be considered for vibration measurement [34, 35].

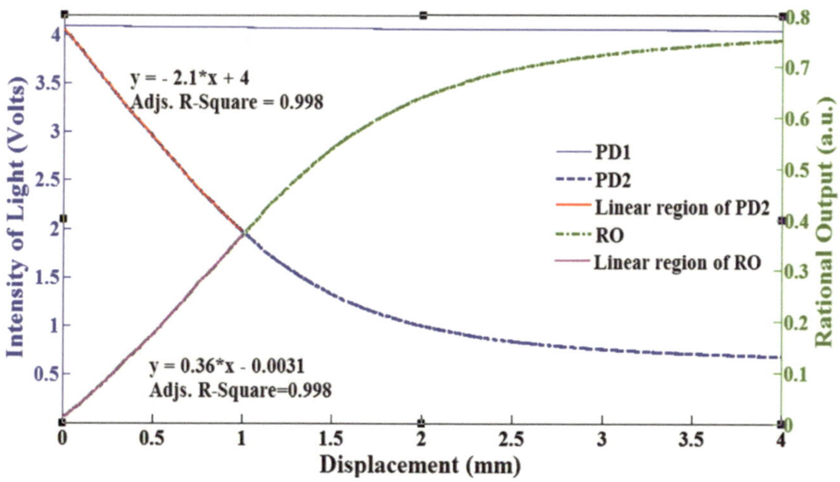

Figure 11.
Displacement characteristic curve of the fiber optic 2x2 coupler vibration sensor.

4.2 Elimination of source and bending fluctuations

Prior to the vibration measurement, the sensor response is tested against source fluctuation and fiber bending at the source end. **Figure 3** shows the effect of source fluctuation on the sensing and reference signals. The measured signals from PD1 and PD2 show a change in intensity of light with respect to variation in light intensity of LED by means of varying the driving voltage, whereas the RO of these signals show insensitive to the source fluctuations. It is apparent from the test results that the RO method minimizes the effect of source fluctuations on the response of the sensor. Similarly, to test the effect of bending losses of optical fiber on the sensor output, the optical fiber is allowed to undergo bending by using a microbending pressure element. **Figure 12** illustrates the effect of fiber microbending at the source end (port2) on individual outputs of PD1 and PD2 as well as on RO of both the signals. It is evident from the test results that even though the outputs of PD1 and PD2 are affected by the fiber bending, the same is not present in the RO [34, 35].

4.3 Vibration measurement setup

The schematic experimental setup of the fiber optic 2x2 fused coupler for the vibration measurement is shown in **Figure 13**. The setup is mounted on a vibration free table. To test the sensor response for corresponds to vibrations, a synthesized function generator and a commercial speaker having dimension of 25 mm depth and diaphragm of 65 mm diameter with a reflector attached at the center of the diaphragm are used. Data acquisition system is employed to record the TDS of the sensor and to monitor the sensor response for known frequencies and amplitudes of vibration of the speaker.

In general, most of the vibrations are sinusoidal displacements of the vibrating object about its mean position. Generally, this nature of vibrations can be detected by measuring its amplitude and frequency only. Thus, the FFT technique have been used for the conversion of the TDS response into frequency domain response to analyze the object vibrations in the form of frequency and also to compute the amplitude. The experimentation is repeated for various frequencies to compute the detectable high frequency and to test the reliability of and also amplitude resolution.

4.4 Results and discussion

In general, the signal to noise ratio (SNR) is well-defined as the ratio between the power strength of the signal and the noise. It can be derived from the following formula [41].

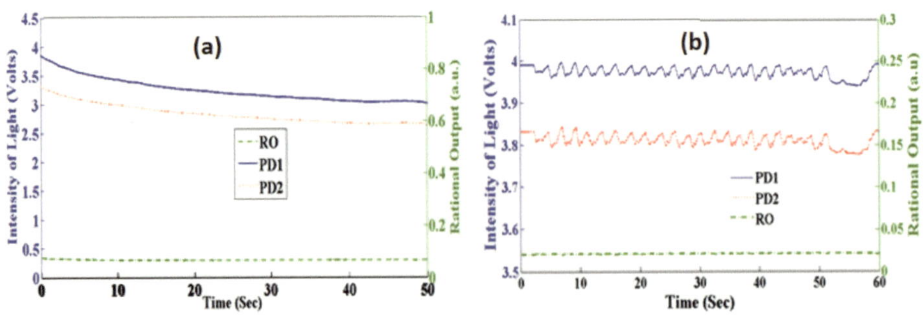

Figure 12.
(a) Effect of source fluctuation on PD1, PD2 and RO, (b) effect of fiber bending on the output of PD1, PD2 and RO.

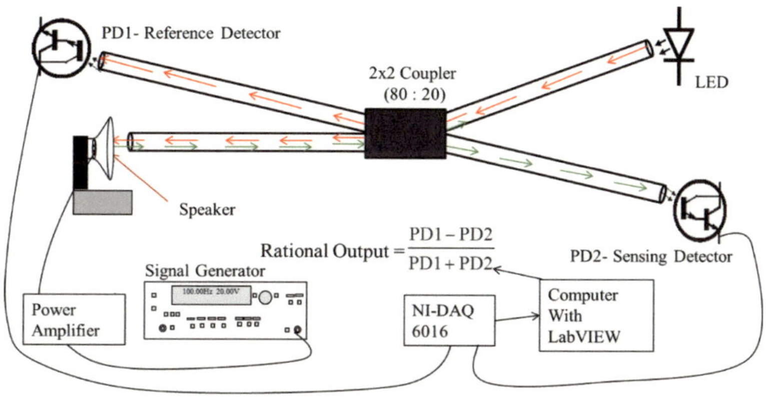

Figure 13.
Schematic of the 2x2 fiber optic vibration sensor experimental setup.

$$\mathrm{SNR} = \frac{\text{Power of signal}}{\text{Power of noise}} = \frac{\mu}{\sigma} \tag{18}$$

Where σ is the standard deviation of the noise signal and μ is the expected value. The SNR is a nominal term which used for the characterization of the quality of the detected signal in a measurement system. In order to measure SNR of the signals of PD1, PD2 and RO, without vibration the sensing probe is maintained at constant distance from the speaker for a period time as shown in **Figure 14(a)**. This reveals the stability of the detection signals. The SNR of the normalized RO, PD1 and PD2 signals is calculated using Eq. (18) and found that the SNR of the RO is improved when compared to PD2 which is clearly shown in **Figure 14(b)**.

To assess the sensor vibration response, a sine wave is applied to the speaker and corresponding TDS response of the sensor is recorded by using DAQ at certain frequency as shown in **Figure 15**. The TDS waveform of RO recorded by the DAQ and corresponding FFT spectrum at 1 kHz signal brings out the closeness with which the sensor responds to a given frequency of applied vibration [34, 35].

The sensor is also tested for its amplitude response by the application of a gated signal of constant frequency, and noting the correspondingPD1, PD2 and RO waveforms as shown in **Figure 16**. It is illustrated that the amplitude of the signal applied to the speaker is constant for a period of 1.1 sec, followed by damped decay of the signal representing that the signal generator is switched off and later by the dc signal, indicating the absence of the signal during this period. This figure also depicts that the output of the vibration sensor at a given frequency the amplitude

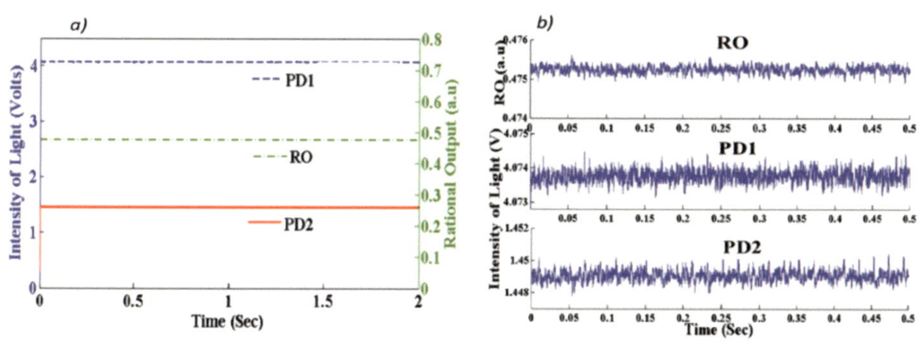

Figure 14.
(a) Stability of the signals and (b) base noise in the signals.

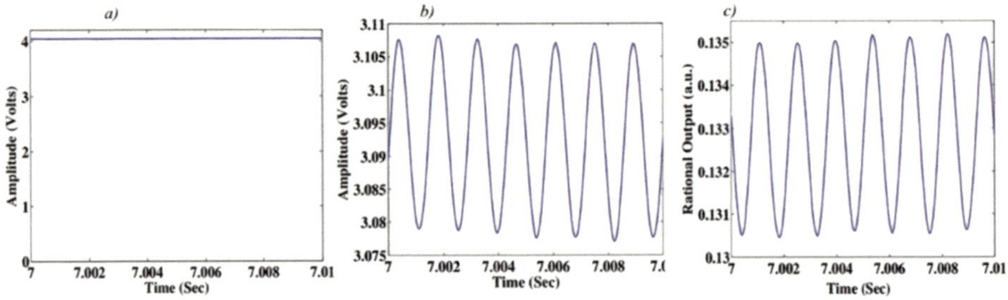

Figure 15.
The TDS waveforms of (a) PD1, (b) PD2 and (c) RO for a frequency of 700 Hz.

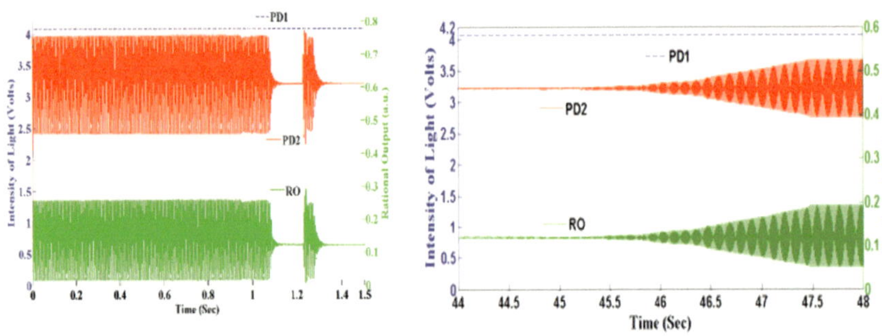

Figure 16.
The amplitude response of the sensor for corresponding speaker vibration at 390 Hz. TDS spectra of the sensor corresponds to continuous variation.

response for the vibration. The amplitude response of the vibration sensor for the applied driving voltage to the vibrator and peak voltage of the FFT from the output signal at for various frequencies observed that the linear in response. The sensor amplitude sensitivity with respect to applied frequency to the speaker exhibits a linear response [34, 35].

From the above reported results, when compared all the responses of different configuration such as dual POF, 1x2 fused coupler POF and 2x2 fused coupler POF vibration sensors, and it was found that among these sensors, 2x2 coupler has been shown better response, which are tabulated in **Table 2**.

Configuration	Linear region (μm)	Range (μm)	Sensitivity (mV/μm)	Linearity	Frequency range (Hz)	Amplitude resolution ~ μm
Dual fiber	600–1300	700	9.83	0.999	900	1
1x2 Coupler	0–1000	1000	2.483	0.9993	1300	1
2x2 Coupler	0–1000	1000	2.1	0.99	3500	0.03

Table 2.
The effect of light source on the vibration response of the sensor.

5. Summary

An all Plastic optical fiber (POF) physically non-contact vibration sensors are discussed, that works based on the reflected light intensity modulation reported with various structures. For every system, observed development and an improvement, with proper design and eliminated the dark region, having a single slope. It enables the system of easy alignment. By considering the rational output measurement, it has eliminated some of the significant effects on sensing such as power fluctuation in light source and also bending losses. When compared the dual POF, POF fused 1x2 coupler and POF fused 2x2 coupler vibration sensors response results. It is clearly predicted that the POF 2x2 fused coupler vibration sensor exhibits enhanced response with 0.03 μm high resolution up to 3.5 kHz frequency range.

Author details

Putha Kishore[1*], Dantala Dinakar[2] and Manchineellu Padmavathi[3]

1 Global College of Engineering and Technology, Kadapa, India

2 National Institute of Technology Warangal, India

3 SKR and SKR Government Degree College for Women (A), Kadapa, AP, India

*Address all correspondence to: kishore.photonics@gmail.com

References

[1] Brain Culshaw and Alan Kersey, "Fiber Optic Sensing: A Historical Perspective," Jour. of Lightwave Tech., vol. 26, no. 9, may 2008.

[2] Eric Udd, "An overview of fiber-optic sensors", *Rev. Sci. Instrum.*, Vol. 66, No. 6, pp. 4015–4030, 1995.

[3] D. A. Krohn, "Fiber Optic Sensors: Fundamentals and Applications," Second Edi., *ISA publisher*, USA, 1992.

[4] Tarun Kumar Gangopadyay, "Review: Prospects for Fiber Bragg Grattings and Fabry-Perot, Interferometers in fiber-optic vibration sensing," *Sensors and Actuators A*, vol. 113, pp. 20–38, 2004.

[5] Yoany Rodriguez Garcia, Jestis M. Corres and Javier Goicoechea, "Review Article: vibration Detection using Optical Fiber Sensors," *Journal of Sensors*, vol. 2010, no. 936487, 2010.

[6] N. Sathitanon,and S. Pullteap, "A Fiber Optic Interferometric Sensor for Dynamic Measurement," International Journal of Computer Science and Engineering, vol.2, no.2, pp. 63–66, 2008.

[7] K. T. V. Grattan, Dr. T. Sun, "Fiber Optic sensor technology: an overview," Sensors and Actuators, vol. 82, pp. 40–61, 2000.

[8] Bahareh Gholamzadeh and Hooman Nabovati, "Fiber Optic Sensors," World Aac. Of Sci. Eng. and Tech., vol. 2, pp. 06–29, 2008.

[9] N. K. Pandey and B. C. Yadav, "Embedded fiber optic microbend sensor for measurement of high pressure and crack detection," Sensors and Actuators A, vol. 128, pp. 33–36, 2006.

[10] Denis Donlagic and Miha Zavrsnik, "Fiber-optic microbend sensor structure," Optics Letters, vol. 22, no. 11, pp. 837–839, 1997.

[11] S. Binu, V. P. Mahadevan Pillai and N. Chandrasekaran, "Fiber Optic displacement sensor for the measurement of amplitude and frequency of vibration", Optics & Laser Technology, Vol. 39, pp.1537–1543, 2007.

[12] M. Yasin, S. W. Harun, Kusminarto and H. Ahmad, "A simple design of vibration sensor using fiber optic displacement sensor," OAM-RC, Vol. 4, No. 1, pp. 1795–1797, 2010.

[13] Guido Perrone and Alberto Vallan, "A Low-Cost Optical Sensor for Noncontact Vibration Measurements," IEEE Trans. On Inst. And Meas., vol. 58, no. 5, pp. 1650–1656, May 2009.

[14] N. E. Lewis, M. B. Miller, W. H. Lewis, "Fiber Optic sensors utilizing surface reflections, in

[15] A. Buffa, G. Perrone, and A. Vallan, "A Plastic Optical Fiber Sensor for Vibration Measurements," in I2MTC 2008 - IEEE International Instrumentation and Measurement Technology Conference Victoria, Vancouver Island, Canada, 2008, pp. 1–5.

[16] M. Linec and D. Donlagic, "A Plastic Optical Fiber Microbend Sensor Used as a Low-Cost Anti-Squeeze Detector," IEEE Sens. J., vol. 7, no. 9, pp. 1262–1267, 2007.

[17] P. B Buchade and A. D Shaligram, "Influence of fiber geometry on the performance of two-fiber displacement sensor," Sensors Actuators A, vol. 136, pp. 199–204, 2007.

[18] M. Yasin, S. W Harun, K. Karyono, and H. Ahmad, "Fiber-Optic Displacement Sensor Using a Multimode Bundle Fiber," Microw. Opt.

Technol. Lett., vol. 50, no. 3, pp. 661–663, 2008.

[19] M. L. Casalicchio, G. Perrone, and A. Vallan, "A Fiber Optic Sensor for Displacement and Acceleration Measurements in Vibration Tests," in IEEE, I2MTC 2009 - International Instrumentation and Measurement Technology Conference Singapore, May 2009, 2009, no. May, pp. 5–7.

[20] P. Kishore, D. Dinakar , D. Sen Gupta, P. Saidi Reddy, M. Sai Shankar, K.Srimannarayana, "Fiber Optic Vibration Sensor Using PMMA Fiber for Real Time Monitoring" Sensors & Transducers journal, Vol.136, Issue 1, pp.50–58, January 2012.

[21] D. Dinakar, P. Kishore, M. Padmavathi, "Study on Intensity Modulated Non-Contact Optical Fiber Vibration Sensors of Different Configuration", Boston USA, Apr 24–25, 2017, 19 (4), pp. 2008–2011, Part XVI, World Academy of Science, Engineering and Technology, International Journal of Physical and Mathematical Sciences Vol:4, No:4, 2017.

[22] K. Imoto, H. Sano, and M. Maeda, "Plastic optical fiber star coupler.," Appl. Opt., vol. 25, no. 19, p. 3443, Oct. 1986.

[23] G. Economou and D. E. N. Davies, "Studies of an optical fibre displacement sensor," J. Inst. Electron. Radio Eng., vol. 57, no. 2, p. 63, 1987.

[24] T. Sugita, T. Abe, K. Hirano, and Y. Itoh, "Bidirectional optical coupler for plastic optical fibers," Appl. Opt., vol. 44, no. 15, pp. 2933–42, May 2005.

[25] H. A. Rahman, S. W. Harun, N. Saidin, M. Yasin, and H. Ahmad, "Fiber Optic Displacement Sensor for Temperature Measurement," IEEE Sens. J., vol. 12, no. 5, pp. 1361–1364, May 2012.

[26] M. Yasin, S. W. Harun, Z. a. Ghani, and H. Ahmad, "Performance comparison between plastic-based fiber bundle and multimode fused coupler as probes in displacement sensors," Laser Phys., vol. 20, no. 10, pp. 1890–1893, Sep. 2010.

[27] A. Wang, M. S. Miller, A. J. Plante, M. F. Gunther, K. A Murphy, and R. O. Claus, "Split-spectrum intensity-based optical fiber sensors for measurement of microdisplacement, strain, and pressure," Appl. Opt., vol. 35, no. 15, pp. 2595–601, May 1996.

[28] M. Yasin, S. W. Harun, and H. Ahmad, "Simple design of optical fiber displacement sensor using a multimode fiber coupler," Laser Phys., vol. 19, no. 7, pp. 1446–1449, Jul. 2009.

[29] Samian, Y. Hadi Pramono, A. Yunus Rohedi, F. RUSYDI, and A. ZAIDAN, "Theoretical and experimental study of fiber-optic displacement sensor using multimode fiber coupler," J. Optoelectr. Biomed. Mate., vol. 1, no. 3, pp. 303–308, 2009.

[30] A. R Crawford and S. Crawford, "The Simplified Handbook of Vibration Analysis Volume 2: Applied Vibration Analysis," Computational Systems, Incorporated (CSI), vol II, 1992.

[31] P. Kishore, D. Dinakar, K. Srimannarayana, P.Vengal Rao, "A simple 1x2 plastic optical fiber coupler based vibration sensor", an International conference on Optics in Precision Engineering and Nano Technology, Singapore Expo, Expo 1 Drive, Singapore, April, 2013, Proc. Of SPIE Vol. 8769, pp. 876929–1 to 876929–4.

[32] P. Kishore, D. Dinakar, P.Vengal Rao, K. Srimannarayana, "Fiber Optic Fused 1x2 Coupler as a Vibration Sensor", Optica Applicata, issue 4, vol. 43, pp. 793–800, 2013.

[33] P. Kishore, D. Dinakar,, K. Srimannarayana and P.Vengal Rao, "A

Study on the effect of light source on Fiber Optic Fused 1x2 Coupler Vibration Sensor", Journal of Optoelectronics And Advanced Materials, Vol. 16, No. 11–12, pp. 1232–1236, Jan-Dec. 2014.

[34] D. Dinakar , P. Kishore, P. Vengal Rao and K.Srimannarayana, "Remote vibration sensor using fiber optic fused 2x2 coupler", Proceeding of SPIE Volume 8836, pp. 883605–1 to 883605–9, Optomechanical Engineering 2013, USA.

[35] P. Kishore, D. Dinakar, P.Vengal Rao, K. Srimannarayana, "Vibration sensor using 2x2 fiber Optic Coupler", Optical Engineering, issue 10, vol. 52, pp. 10704, 2013.

[36] A. Djordjevich and S. Savovic, "Investigation of mode coupling in step index plastic optical fibers using the power flow equation," IEEE Photonics Technol. Lett., vol. 12, no. 11, pp. 1489–1491, Nov. 2000.

[37] J. Chang, Q. Wang, X. Zhang, L. Ma, T. Liu, and Z. Liu, "Single-end vibration sensor based on an over-coupled fiber-loop reflector," Laser Phys., vol. 18, no. 4, pp. 452–454, 2008.

[38] J. M. Lourenço and P. M. Cavaleiro, "Data processing for intensity based fiber optic sensors," Appl. Opt., vol. 35, no. 34, pp. 6835–6, Dec. 1996.

[39] X. Jian and M. Bin, "Experimental Investigation of Coupling Fiber-Optic Sensor for Vibration Measurement," International Conference on Electrical and Control Engineering, pp. 936–939, 2010.

[40] M. Endong, W. Min, P. Fufei, and W. Tingyun, "Fiber Optic Coupler as Sensing Probe of Temperature Sensor," 8th International Conference on Electronic Measurement and Instruments, pp. 4–186–4–189, 2007.

[41] Daniel J. Schroeder, "Astronomical Optics," Academic Press, Second Edition, San Diego, pp. 433–435, 2000.

Section 2

Optoelectronics Materials and Structures

Chapter 5

Periodic Nanophotonic Structures-Based Light Management for Solar Energy Harvesting

Nikhil Deep Gupta

Abstract

Solar energy has always been an obvious choice for solving the energy issues for the humans for centuries. The two most popular choices, out of many, to harness this infinite source of energy are: solar cells and photoelectrochemical cells. Although both these techniques are quite attractive, they have inherent limitations for tapping all of the incident photons. Maximizing the absorption of incident photons to produce maximum possible electrical output is always the main impetus for the researchers working to streamline these two techniques and making them compatible with existing sources of electrical energy. It has been well established that the light trapping in the solar cells and photoelectrochemical cells can play a vital role in improving their performance. To design light harvesting structures for both these applications, periodic nanophotonic structures have demonstrated stupendous results and shown that they have the real potential to enhance their performance. The chapter, in this regard, presents and reviews the current and historical aspects of the light harvesting structures for these two interesting applications and also discusses about the future of the research to further the performance of these large-area solar-to-electrical conversion transducers.

Keywords: periodic nanophotonic structures, solar cells, photoelectrochemical cells, photonic band gap, diffraction grating

1. Introduction

As per the United Nations Foundation report [1], one of the major crises the world currently is facing is the climate change and sustainable energy solution. For sustainable development and to curb climate-related problems and at the same time for catering to the ever-increasing energy demands of the humans, more and more countries are shifting toward renewable energy sources. The factor, in turn, drives the development for the future access to affordable and sustainable energy through investment in the efficient renewable energy programs, such as Jawaharlal Nehru National Solar Mission program [2] in India, which is one of the largest programs currently running in the world for renewable energy expansion.

Most part of the world is poised to have the solar energy available and that too for maximum part of the year. Also, it is an infinite source of energy that is freely available and that too without causing any environment adversaries. The energy

IntechOpen

from the sun plays the most important role in the sustenance of life on the earth as almost all the energy sources available here are inherent from the heat and light of the sun itself, either directly or indirectly. The solar energy is being in use from thousands of years through various ways. Still, this vast source of energy is underutilized and, thus, novel and innovative practices are required to harness this unending source of energy. The total solar energy received on the surface of the earth is accounted to about 3,850,000 EJ/year [3] and it is more than twice the total energy that can be ever gained from all of the earth's nonrenewable energy resources combined. Hence, harnessing the solar energy for the betterment of the society has always attracted the researchers.

To tap the solar energy, many applications are being done, starting from the solar thermal innovations to the solar photovoltaics applications. Out of them, direct conversion of the solar incidence to the electrical energy is one of the prominent reasons for the immense increase in the popularity of the solar energy related applications. Two of the most prominent techniques that directly convert the solar energy into the electrical energy are the solar cells (SCs) and the photoelectrochemical cells (PECs). The research and innovations in these optoelectronic devices have regained interest during the last two decades or so, on account of their proven capability to effectively harness the solar energy. The task is to make them compatible and competitive with the available energy resources.

A simplistic design of a complete SC is shown in **Figure 1**. The principle of operation of SC is well known for years [4], and they can work satisfactory for long without much requirement of maintenance. The operation of a SC can be summed up in three processes that include carrier generation upon incidence of light, carrier separation, and carrier collection at the outer electrodes [4]. The generation of charge carriers occurs in the active region, which is of the materials, which on absorption of a photon generate electron-hole pairs. The absorption in the active material is only responsible for the useful charge carrier generation. These generated charge carriers are then separated because of a built-in electric field provided through designs such as p-n junction and then are effectively collected at the outer electrodes.

The reflections from the top account for more than 30% of the losses from a SC. To reduce these reflections, an antireflection coating (ARC) is placed at the top as an integral part of the design of a SC. On the other front, to reduce the transmission of the unabsorbed photons from the active layer (AL), back reflectors (BRs) are being employed to reflect back these unattended photons. Commonly, both ARC and BR are considered as a part of light management schemes (LMSs) for the SC. Although, the incorporation of LMS increases the cost and complexity of the overall device, the performance enhancement that they provide helps to reduce

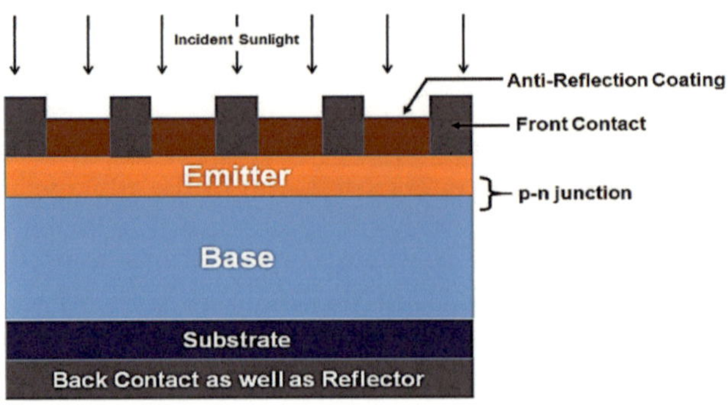

Figure 1.
A basic structure of a solar cell.

the final cost-to-efficiency (CE) ratio of these devices. The LMS structures play a very critical role in the SC designs and are becoming even more crucial with the reducing thickness of the ALs in the novel designs.

On the other hand, solar-to-electricity conversion through chemical route using PEC process is also becoming significant due to its capability of long-term energy storage without adversely affecting the environment. PEC uses to generate hydrogen through water splitting using solar energy, also commonly known as photo-electrolysis. The created free hydrogen due to photo-electrolysis can then be utilized for various applications. In a commonly used PEC design, there are two electrodes, namely, a working photo-anode and a counter metal-electrode as shown in **Figure 2** (although three electrodes' design is also being used, with the third electrode known as reference electrode). The photo-anode acts as a photocatalyst that absorbs the photons from the solar flux incident on it and creates charge carriers. Then, these carriers move to the opposite polarity electrodes that leads to the generation of the electrical energy.

In both of the abovementioned energy harvesting operations using optoelectronic devices, the main aspect that one looks toward to define their quantitative performance is their photo-conversion efficiency (PCE). The recent past has observed a considerable improvement in the PCE for these devices. In the case of SC, this pace of increase in the PCE is even faster as compared to PEC. For a crystalline Si-SC, the efficiency has almost reached to the limiting values [5, 6], whereas for PEC the efficiency is well under 20% [7]. However, there is still a gap that is required to be filled before these devices can certainly overtake their conventional energy counterparts. One of the most important parameters that affects these devices' mass acceptance is their CE ratio.

To reduce the CE ratio, it is obvious that either one has to reduce the cost of design or increase the PCE, or one can target both. One of the ways to reduce cost is to go for thin film technologies for design. However, it has a trade-off as such designs adversely affect the PCE. In order to compensate for the losses incurred, one of the time-tested methods is the use of LMS. The chapter is devoted to the discussions related to the design and innovations for the LMS for these solar energy harvesting optoelectronic devices. The LMS makes it possible for both SC and PCE structures to notably improve their PCE. Nanophotonics structures are playing a very crucial and successful role in designing LMS, and it has been demonstrated through several studies that they have a realistic potential to make these devices to work at their limiting values or can even go beyond that.

The chapter is organized as follows: Section 2 discusses about the need for LMS for solar optoelectronic devices. The role that periodic nanophotonic structures can play is discussed in Section 3. That section also deals with the recent advancements

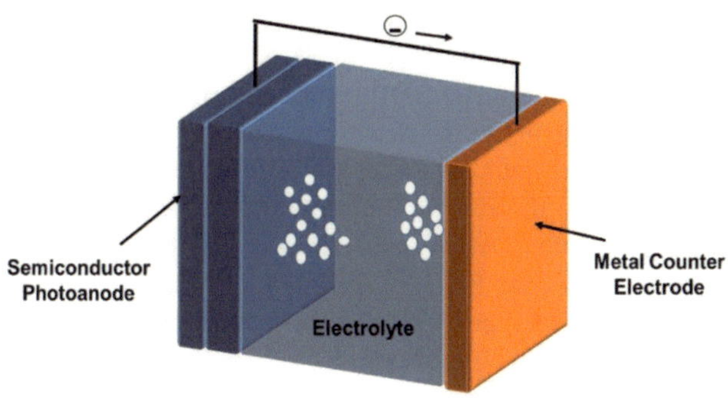

Figure 2.
A basic structure of a photoelectrochemical cell with two electrodes.

and innovations related to LMS in the field of SC and PEC, respectively. Section 4 discusses the present challenges and future perspective related to the mentioned optoelectronics devices followed by concluding remarks.

2. Need of light management for solar optoelectronic devices

From past few decades, it has been demonstrated that the thin film (TF) technologies have the reasonable capability to perform the operation of optoelectronics solar energy harvester (OSEH) with an additional advantages of low cost, lightweight, flexibility, comparatively easier fabrication processing, and higher production throughput [8]. To make the CE ratio of these OSEH comparable with the existing technologies, the ALs' thicknesses used in these devices are becoming thinner and thinner, as it can check the design cost. Although, the reduction in the thickness of ALs control the overall design cost, the final device has to compromise with the PCE, as with reduction in thickness, the number of photons that can be absorbed in an AL are also get restricted [4, 9]. The restriction is due to the fact that each material has its own absorption coefficient that dictates the incident photons' absorption within its layer's specified thickness, and if the material thickness is lesser than that, the photon remains unattended and can pass through the device unabsorbed without any useful contribution. In other words, the photon gets wasted in such a case.

Hence, a large part of the incident spectrum has wasted because of this incomplete absorption. Together with the loss due to limited thickness of the AL, there are also major losses due to reflections from the top surface as well as due to recombination, including others. In such circumstances, to curb these losses and increase the useful absorption of the otherwise lost photons is utmost required to improve PCE of these devices. To enhance the absorption, the most practical and profoundly used technique is light management structure. LMS is the process to effectively couple the incident photon to the AL and make it possible for the layer to successfully absorb it and contribute to the useful charge carrier generation. As the focus of the LMS is to make it possible for an incident photon to be effectively get absorbed within the active material, LMS can be designed in combinations with the ALs in various ways depending upon the material used and the application requirement [10, 11].

The use of ARC at the top of the SC and PEC designs is a part of a LMS, where the main task of ARC is to reduce reflections from the top that arises due to the sudden variation of the refractive index from the incident medium to the absorbing medium. ARC, thus, requires to serve the purpose of index matching for the incident light with the absorbing layers and increase useful absorption of photons [4, 9]. ARC in a simpler design can be carved out as a single planar layer whereas its performance can be improved with the incorporation of multiple layers to provide gradual change in refractive index [12, 13]. In all, the ARC structure must have negligible reflections with insignificant absorption for the incident spectrum. However, practically observing such a property out of the planar ARC design that can achieve supreme coupling without losses throughout the incident spectrum, and ultimately restricts the outward movement of coupled photons from it through oblique escape angles, is really challenging [9]. For planar ARC, there is a need to appropriately find an optimized central wavelength, λ_c, to get relatively low reflections around the prioritized wavelength range. A planar ARC has almost zero reflections at λ_c.

The thickness of the planar ARC design (T_{ARC}) can be optimized with value of λ_c as it is equal to the quarter of the central wavelength. Considering n_{ARC} as the refractive index of the single planar ARC layer, mathematically T_{ARC} is given as

$$T_{ARC} = \frac{\lambda_c}{4n_{ARC}} \tag{1}$$

In contrast, LMS when design at the back of the ALs or when ALs in itself are being used as the LMS, they are intended for increasing the optical path length of the coupled incident photons, virtually through diffraction or scattering in the desired directions, so that they can be absorbed within the limited AL thickness [14, 15]. In such cases, the LMS design is focused for trapping photons that require absorption depth more than the AL thickness to be successfully absorbed in it. The absorption depth requires to tap a photon of certain wavelength depends upon the absorption coefficient of the active material used [4]. The relation between the absorption coefficient, α, and incident photon wavelength, λ is given as:

$$\alpha = {4\pi k}/{\lambda} \tag{2}$$

Here, *k* denotes the materials' refractive index' imaginary part. The absorption depth requires to absorb a particular wavelength photon is given by the inverse of the absorption coefficient. It is clear from the Eq. (2) that as the incident photon wavelength increases, the absorption coefficient decreases and in turn comparatively thicker layers are required to absorb these photons. So, materials with high absorption coefficients should be more preferred for better absorption of incident photons with less thickness. However, for keeping the CE ratio to a lowest possible side and also for creating flexible structures, the thickness of the AL is always kept less than the required thickness for absorbing the possible highest wavelength photons. The case is even worse for Si, the most preferred material for designing SC, as it has very low α values near to its band edge.

In the case of PEC designs, the LMS has an another important role to play. For PEC devices, from last decade or so, Group-III nitride materials, such as $In_xGa_{1-x}N$, have emerged as the leading source for designing photo-electrodes. It is due to the possibility to tune their bandgap over the large wavelength range (from 3.4 to 0.65 eV) (depending upon the concentration of Indium in GaN) [16–18]. With the possibility of tuning the bandgap, the $In_xGa_{1-x}N$ materials seems to have the potential to provide full spectrum operations. Other favorable properties of group-III nitride materials for PEC devices include sufficiently good irradiation resistance chemical tolerance, thermal stability, carrier mobility, direct bandgap property with significantly high absorption coefficient even near to the band edge that make them suitable and a preferred choice over metal oxides for PEC operations [19–21].

Although, $In_xGa_{1-x}N$ seems to be a best fit for the PEC devices, the main limitation is the thickness of its good quality epitaxial film that can be grown over the substrate. This remains one of the unsolved queries for $In_xGa_{1-x}N$, because of which even with many favorable optoelectronics properties, still $In_xGa_{1-x}N$-based designs are not prevalent especially for SC applications. $In_xGa_{1-x}N$ material is basically grown on different kind of substrates that always has a lattice mismatch (sapphire is typically the preferred substrate material for growing III-nitride materials) [22]. To tune the bandgap of $In_xGa_{1-x}N$ to absorb maximum incident solar spectrum, it is required to increase the In incorporation, but with increase in In composition, the quality of grown film degrades. This is due to the increase in defect densities with increase in In incorporation, and in turn, puts a halt on the PCE possible from the $In_xGa_{1-x}N$-based devices [23]. In such a case, LMS as a BR becomes crucial, as it is a support structure that can make the design to achieve high absorption even with limited thickness of active layer with limited In content.

Designing an ideal low loss BR with a planar metallic layer [24] that can perform equal reflections throughout the desired range of wavelengths is a tedious task and practically is not able to achieve even half of the limiting values [25]. Theoretically, it is possible to achieve an enhancement of $4n^2$ in absorption with an ideal LMS (where n denotes refractive index of the active material) [25]. However, with the planar LMS design the limiting values has not been achieved till date because of several factors such as the intrinsic losses from surface plasmon modes generated at the granular metal-dielectric interface [26]. Due to the inherent limitations in absorption enhancement that planar LMS achieve, it becomes imperative for the researchers to look toward the design of low loss, near-to-ideal LMS that can let the development of the OSEH that can work near to the limiting values [6] to get maximum output from the incident sunlight. The possible solution lies in the use of the nanophotonic structures. They have regularly demonstrated their potential for the design of LMS and will be discussed in detail in next section.

3. Periodic nanophotonic structures for light trapping

Now, as it is well established fact that with the commonly used planar ARC and BR structures, it is not ever possible to achieve the limiting PCE values, so it is very much required to look for the better and efficient alternatives to design LMS that can include both random as well as periodic nanostructures. Although, the random structures are easy to fabricate and design with minimum infrastructure requirement, they are complex to reproduce and rescale for the industry required bulk production, it is better to look for the periodic structures for designing LMS. Periodic nanophotonic structures (PNS) having subwavelength dimensions have been utilized and demonstrated their effectiveness successfully for a variety of optoelectronic applications [27–31]. PNS have a lot to offer for the design of optoelectronic devices as they have various unique advantages especially for the OSEH applications. They have been successfully utilized and demonstrated their superiority for designing LMS.

These PNS have the better capability to couple the incident photons to the AL as compared to the planar LMS and thus can help to enhance the charge carrier generation, surface to volume ratio that can boost the quantum effects, reduce recombination that accounts to major losses, enhance carrier collection at electrodes, provide tunable bandgap property etc. [32, 33]. The PNS can be coupled with any of the active material system in use and contribute according to their properties. With unique properties of PNS, one can have the advantage of manipulating the light propagation and light-matter interaction as per requirement and opens up the wide range of possibilities in variety of applications for the field of optics and photonics, including OSEH applications.

These structures can be incorporated in the various ways within the OSEH devices, as per the particular application's requirement. Researchers have examined PNS with OSEH in different forms extending from using them as diffraction element to the back reflector or even carving active material itself as PNS to enhance the absorption of incident photons. The PNS are mainly responsible for taking the PCE of these energy harvesters near to the limiting values or in some cases even surpassed them. In the coming subsections, the article will discuss in detail about the history and advancement that has been done in the field of PNS-based optoelectronics solar energy harvesters, especially SC and PEC.

3.1 Light management through periodic nanophotonic structures in solar cells

Since, the beginning of the current century, the research and development in the field of PNS-based SC designs have seen an upward swing, irrespective of the active material used, starting from Si to perovskites. The fact can be recognized with the exponential increase in the number of research articles that have emerged out of the area. The researchers have conducted studies for the amalgamations of PNS with the SC from different perspectives, as they used them as a single low loss dielectric BR [34], as a diffraction grating [14, 35], designing the absorbing material itself as PNS [36, 37]. These studies are basically performed either using 1D or 2D or pseudo 3D PNS [38–41], as designing and fabrication of 3D PNS is still a challenging task.

In one of the interesting findings, Wang et al. in [42], demonstrated the critical role that PNS has to play for the improvement in the performance of thin film SC and their studies was based on GaAs active material. It is important to highlight the findings because GaAs has better absorption coefficient and radiative recombination dominates as compared to Si and earlier it was thought that the BR could only be important for materials such as Si having less band edge absorption coefficient [25]. In case of GaAs, specially, the BR becomes important to reflect back the unabsorbed photons that are radiatively emitted and let them to contribute through photon recycling. Following the proposal studies, Gupta et.al in [15] proposed a PNS-based GaAs structure, as shown in **Figure 3**, having only 500 nm active layer thickness. They have presented that the PNS having a combination of 2D and 1D structures to provide a pseudo 3D PNS has a capability to provide enhancement of about 46% as compared to the planar LMS design.

Another important point that they presented was that the effect of BR decreases with increase in the AL thickness. They have demonstrated that the enhancement contribution from PNS structures for PCE was about 200% in case of 100-nm active layer cell that reduces to only around 25% for 1-μm thick AL cell. Thus, for thicker cells, the BR role is limited to only near to the band edge wavelengths at most and in such cases, even planar layers can accomplish the task.

Zhou et al. illustrated one of the classical findings for PNS in [14]. They have used a-Si:H as the active material and designed LMS for the same. In the LMS, they have utilized 1D PNS at the back of the SC as distributed Bragg reflectors (DBRs), and in between the active layer and DBR, there was a layer of 2D PNS placed to enhance absorption through diffraction of light at oblique angles. The authors put forth the physical insight of the whole mechanism and emphasized an important condition that the PNS lattice constant must be comparable to the wavelength of light in the medium, which it meant to manipulate.

In another interesting study, Chutinan et al. in [43], for thin crystalline Si-SC, demonstrated that it is possible to notably enhance the device PCE with PNS.

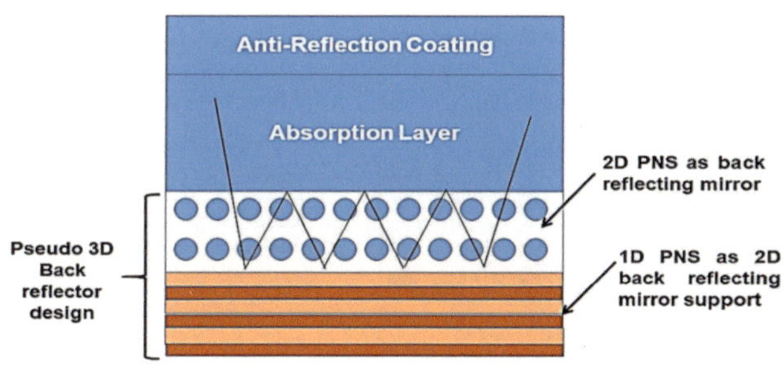

Figure 3.
Schematic design of a SC using pseudo 3D PNS as back reflector.

They have demonstrated the comparative enhancement of around 11 and 3% for 2 and 10 μm active layer thick SC designs, respectively, for Si AL. The SC design was achieved through carving active layer itself as PNS, as shown in **Figure 4**. The article demonstrated a decent physical interpretation of the SC design having PNS-based LMS within absorbing layer. As the PNS can be equally implemented for other material systems, a work has recently also demonstrated the effect of PNS within active layer for perovskite-based design [44]. The results have shown that the PNS are also critical for the design of perovskites-based SC as the defects in the deposited perovskites films becomes prominent with their increasing thickness.

In one of the exciting studies, Munday in [45] presented and analyzed the Si-SC design having LMS targeted from the top coupling surface using the PNS. The PNS was designed to work as Photonic Band Gap (PBG) structure. The PBG is a kind of structure that has a forbidden band for the certain range of frequencies and does not allow these forbidden frequencies to pass through them [38]. The PBG at the top was intended in the design to block an absorption of a range of incident photons but also intended to disallow the same range of photons' emissions. The authors through the design claimed that the PBG placed at the top would reduce the spectrum available for absorption but at the same time through stopping the emission out of the SC, one could achieve higher *Voc* without effecting the *Isc* from the SC. The design increases the overall PCE as the minority carriers' density available for absorption increases. The PBG influence can be more effective for designs using direct band gap materials, such as GaAs. In direct band gap materials, radiative recombination dominates and photons can be recycled after it is released due to radiative recombination and can be reused. In such cases, because of PBG structure at top, the recycled photon will not be allowed to direct out of the SC and would be available for absorption again within the SC.

Bozzola et al. in [36] and Zannoto et al. in [37] also demonstrated Si-SC designs with PNS-based LMS at the top surface facing the incident light, as shown in **Figure 5**. However, in contrast to Munday, they have used PNS as the diffraction grating to maximize the coupling of the incident solar spectrum to the AL through reduction in reflection and creating gradual refractive index variations. For PNS-based diffraction grating designs, it is required that the plane of periodicity must be perpendicular to the incident light. The design, instead of targeting or filtering only a range of wavelengths, has to work for the entire solar spectrum, and thus, needs to be carefully optimized.

In some other studies, the researchers have also demonstrated the designs for the PNS-based diffraction gratings at the top for different materials such as GaAs [9], InGaN/GaN [18] active materials. The results in all the cases, irrespective of the materials used have shown that these PNS have immense potential for taking the performance of the SC to new heights.

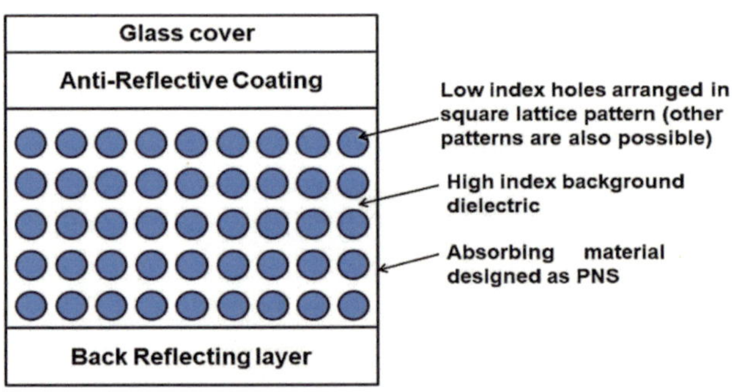

Figure 4.
Schematic design of a SC using PNS as absorbing layer.

In the article presented by Demesy et al. [46], it was discussed that it is possible to achieve PCE in the range of 15–20% just with 1 μm of Si active layer, if they are designed in the form of modulated nanowires. With careful optimization of the modulation profile, one can achieve significant enhancement in antireflection property, better light management, and back reflections of high wavelength photons simultaneously over broad angles. In this context, it is really interesting to find the structures that can use two different kinds of PNS structures together in the SC designs such as PhC and nanowires, as shown in **Figure 6**.

In another important study, Mallick et.al [47] also referred that for materials having indirect band gap and low absorption coefficient near the band edge, LMS are of utmost importance for the applications of SC. They have performed the analysis on the 400 nm thick Si AL-based SC. They have optimized the PNS through tuning of coupling photons of particular wavelengths to quasi-guided modes over a broad spectral range. The structure consists of two layers of PNS with different dimensions. The upper layer has a smaller radius of holes as compared to beneath layer. Their analysis has shown that there is a possibility of 8-fold increase in the average photon absorption compared to the planar SC with AL of same volume.

In another work, Eyderman et al. in [48] performed the study of the PNS effect on highly absorbing structures such as GaAs. They have used a slanted conical pore PNS packaged with SiO_2 and deposited on a silver back-reflector for their SC. They have performed the studies with ultra-thin GaAs layers from 100 to 300 nm and demonstrated that it is possible to tap almost 90% of the incident photons in the high wavelength range of 400–860 nm and achieved short circuit current density over 26.3 mA cm^{-2}. The quantitative analysis was further by Gupta et.al in [13] and provided the insight into the importance of the use of PNS-LMS for GaAs, including its effectiveness to improve the angular performance of the device.

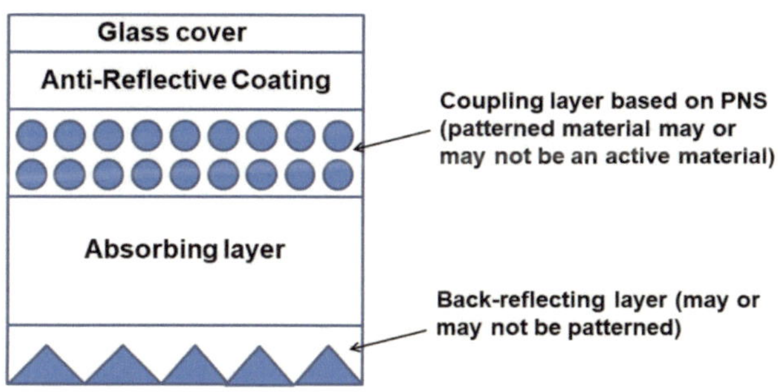

Figure 5.
Schematic design of a SC using PNS at the top for effective coupling of incident light. ARC can also be carved out as PNS structure.

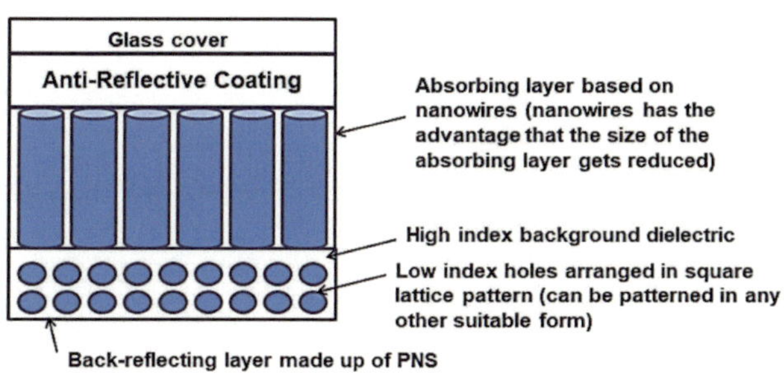

Figure 6.
Schematic design of a SC using nanowires as active structure in combination with PNS-based BR structure.

In one of the recent studies, Peer et al. in [49] has demonstrated theoretically that microlens arrays-based LMS can effectively enhance PCE of perovskite SC, and experimentally showed that the micro-lens can enhance organic SC efficiency. They have fabricated microlenses through nanoimprint lithography at the top surface and thus the design did not impact the material quality of the inside layers. Microlens pitch values near 1 micron were studied and PCE gains of 6% for thick perovskites SC were shown. The gains for thinner AL SC are expected to be larger, and it is conceivable that larger pitch values of several microns may provide better efficiency enhancement.

Bhattacharya et al. [50] and Hseih et al. [51] in their breakthrough research demonstrated that it is even possible to achieve PCE well beyond the Lambertian limits [25, 36] using PNS-LMS. They have demonstrated both theoretically as well as experimentally that using inverted pyramid and Teepee PNS it is possible with Si to surpass the limiting values. The inverted pyramid structure is recreated in **Figure 7**. They claimed that the main reason for overcoming the Si-SC limiting values using around 10–15-μm thick Si active layers are the existence of long lifetime, slow-light resonances, parallel-to-interface refraction and their coupling with external plane waves. These phenomena are not possible to be predicted using ray-optics models. They have demonstrated absorption beyond the limits in the weakly absorbing region of Si, near infrared wavelength range from 950 to 1200 nm. They achieved short circuit current density well beyond 41 mA/cm2. The study can pave the path for future studies related to PNS-based LMS and can have a long-lasting impact.

Till now, we have discussed about the role of PNS-based LMS for the single junction SC. However, it has also been shown that LMS also has a significant role to play for tandem SC designs. In the tandem cells, there are usually two cells (top and bottom) that are series connected electrically and hence it is required that these cells must be current matched. In the adverse case, the device output is limited by the smaller of the two cells' current and thus put the limit on the current of the entire tandem cell. To achieve optimized output, it is thus needed to optimize the thicknesses of each cell to get maximized PCE [52]. In such cases, usually LMS is critical to enhance the current out of the cell having lower current output.

In this regard, Mutitu et al. in [53] demonstrated a PNS-LMS that can be applied to both single junction and tandem cells. The 1D PNS structures are used as band-pass filters at the interface of two cells to reflect low wavelength photons (400–1100 nm) toward the top cell and transmit high wavelength photons. In addition,

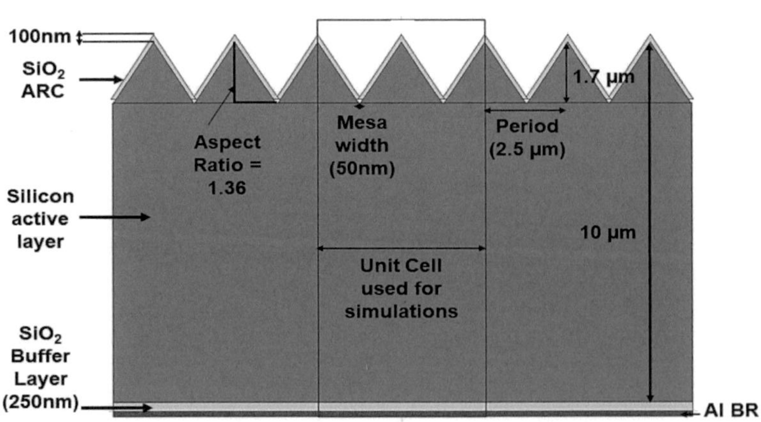

Figure 7.
The schematic diagram of the structure proposed in [51].

nanostructured diffractive gratings were incorporated to redirect incoming waves and hence increase the optical path length of light within the solar cells, and in turn, this has shown significant advancement in the PCE. In another well thought study, the researchers has highlighted the importance of PNS-LMS as intermediate reflectors for tandem cells [52], as shown in **Figure 8**, and achieved significant improvement in the output of the proposed design as compared to planar reflectors.

In the recent past, there is significant development has been done in tandem cells designs using perovskite/Si materials, where higher bandgap perovskite materials are used as the top cell and lower bandgap Si is used as bottom cell to take the final PCE routinely well beyond 30% [54, 55]. Several studies have been proposed in this regard including [56, 57], which have effectively demonstrated the role of PNS-based LMS with perovskite/Si materials tandem SC and help to establish the fact that LMS are also critical for the design of tandem cells irrespective of material system used.

All the abovementioned works, including others not mentioned here due to space limitations, have gradually helped to excavate the insights of the role that PNS-based LMS can play for the SC devices, especially for thin film SC. However, there still lies the challenges and whenever there is a challenge, lies an opportunity. The next section will discuss that in detail regarding the opportunities that are available for the future advancement to make these devices widely acknowledged and adapt more toward designing large area SC.

3.2 Light management for photoelectrochemical cells using periodic nanophotonic structures

PNS-based LMSs have shown above expectation utility for different optoelectronic applications, as already mentioned, and their effectiveness for SC application has already been discussed, and it is apparent to expect an enhanced performance from their adaptation for PEC photo-electrode designs. Various PNS-based LMSs have been studied for designing photo-electrodes, to be used for hydrogen production through water splitting, including, quantum wells [58], photonic crystals [59], quantum dots [60], carbon nanotubes [61], nanowires [62], etc. These PNS-based LMSs have, by one or other, shown their effectiveness for enhancing the photocatalytic activity through enhancing the absorption of incident photons and certainly led to increase in PCE for PECs.

Most of the efforts for PEC photo-electrode designs using PNS-based LMS are intended to modulate the band gap of the structure and take it near to 550 nm to utilize the maximum of the incident photons that have the capability of performing

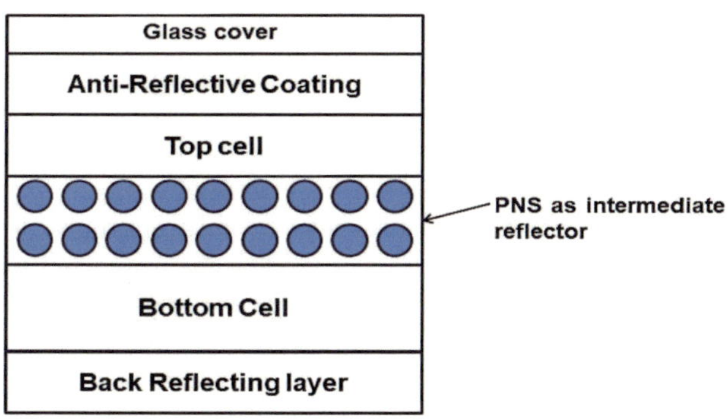

Figure 8.
The schematic diagram of the tandem SC with PNS-based LMS used as intermediate reflector.

photocatalytic action. In this regard, the PEC designs, especially, with nanowires are widely popular and researched. In one of such studies shown by Wang et.al [63], the GaN nanowires were designed using molecular beam epitaxy (MBE) process for wafer-level water splitting. Another highlight of the study was the observation of the stable operation of the PEC system emphasized that the PNS-based photo-electrode design based processes are equally stable in the aqueous solution. Thus, the design can serve the purpose of achieving absorption of the incident photons over the wider spectrum for enhanced photocatalytic operation with stable operation. Such early studies, including [64, 65], have also established the fact that Group-III nitride materials could be a better choice for photo-anode activity for oxygen evolution reaction.

Another study using PNS with Group-III material in [66] has shown that nonpolar GaN could lead to spontaneous water-splitting action together with enhanced proton diffusion process that can be achieved at low energy barrier. Although, the study demonstrated several important aspects of PNS-based PEC operations including achieving demonstration of better stability of GaN-PNS design as compared to TiO2 or ZnO-based design [67], the PCE achieved was still well under the acceptable 10%. The main reason was posed by the bandgap of GaN materials and its absorption coefficient [27, 68]. In another study, the GaN structure designed as nanowire is defect engineered with Mg impurity using epitaxial growth process [69]. In another research, Park et al. in [70], designed metal-assisted GaN nanowires with vapor-liquid-solid process and has done the growth over the graphene film. The structure was then transferred to polymer substrate to create a flexible PEC system. In contrast to GaN, $In_xGa_{1-x}N$ materials, during the past few years have established itself as a better prospect for designing photo-electrode for PECs due to the possibility of better band management with it and thus with PNS they are better positioned theoretically to tap the maximum possible solar irradiation [70, 71].

Kibria et al. were demonstrated the pioneering work on the multiband nanophotonic structures using $In_xGa_{1-x}N$/GaN materials [72]. The design is something kind of a tandem structure designed through carefully controlling the In concentration to do band engineering during the nanostructures growth. They observed the stable PEC operation of the structure till 560 nm, but still the output PCE was limited to 2% only due to the lattice mismatch arises due to the growth of $In_xGa_{1-x}N$ over the unmatched substrate. The researchers tried to further the performance of $In_xGa_{1-x}N$-based PEC operations through the use of light sensitive dyes with nanostructured photo-electrode designs [73, 74], however, the uncontrolled behavior of the space charge properties adversely limited the redox reactions and certainly restricts the PCE.

To improve the carrier extraction process, researchers have observed that controlling the band bending process in nanostructured $In_xGa_{1-x}N$ photo-electrode design and optimizing it is of utmost importance. Such a design has shown a considerable absorption improvement in the PEC process as compared to undoped samples, especially in UV and violet spectrum region [75]. In another study, Alvi et al. achieved In incorporation beyond 40% in the design of nanostructured $In_xGa_{1-x}N$ using PAMBE method and achieved notable enhancement in photocatalytic activity [76]. Caccamo et al. also demonstrated the improved water splitting activity through single crystal $In_{0.3}Ga_{0.7}N$/GaN core-shell type nanowires designed using MOVPE method [77]. In another interesting case, researchers observed experimentally that $In_xGa_{1-x}N$/GaN materials' photo-electrode designed as coaxial multi-quantum well nanowires for PEC operation significantly achieved absorption enhancement and demonstrated final PCE of 8.6% [78]. In one of the important studies, it has been shown that photocatalytic action can also be improved through a core-shell nanostructure-based photo-electrode designed with the combination of Si with $In_xGa_{1-x}N$ as compared to $In_xGa_{1-x}N$ nanostructures grown on Si substrate [79]. Although, inherently, due to lattice mismatch, growing $In_xGa_{1-x}N$ with Si, obviously increases the defects, the use of Si substrate or Si core can offer a benefit from economic perspective, thus need to be explored further.

To further the PCE for the hydrogen production through PEC process, multi-junction photo-electrode designs were also employed as demonstrated by Young et.al [80]. Their tandem structure was designed using GaInP/GaInAs materials in such a way that each junction bandgap could be independently varied. The design has shown a realistic potential to further the PCE and seems to be an exciting perspective. In another recent study, researchers in [81] have achieved the PCE of 8.75% with nanopatterned multiband $In_xGa_{1-x}N$/GaN-based photo-electrode design for one-step pure water splitting system. It has been highlighted in some other studies that for protection of the nanopatterned Group-III nitride photo-electrodes against photo-corrosion and surface oxidation, it is required to make their surfaces nitrogen rich, which ultimately leads to the longer operational life time for these PEC systems [82, 83].

In contrast to nanowire-based designs, Steiner et al. [58] presented the PEC design using quantum well-based LMS with superior strain management for the GaInP/GaAs photo-anode-based tandem PEC device. They highlighted that with tandem structure supported by PNS-based LMS, it is realistic to achieve PCE for PEC beyond the magical 10% value. For semiconductor photo electrode device constructed with quantum wells that can remain stick on the growth substrate, the authors claimed that the proposed configuration can also improve the stability of the system. Still, for such a design, the most challenging issue is to maintain critical strain balance and to reproduce such a structure with the equally desirable strain management must be carefully addressed. The factor ultimately put restriction to further increase the thickness of the active layer and limits the output and one possible way to improve the output further, again, is the better LMS.

In another exciting study, researchers in [59] observed the improved photocatalytic activity using PNS-based LMS accounted due to the enhanced absorption of incident photons near to the band edge of absorption material. They have designed the PNS using titania inverse opal topology. The photocatalytic materials used for photo-electrode was developed using nanoparticles of CeO2, ZrO2, and Y2O3. They observed that the PNS design's period has to be carefully optimized to enhance the absorption of the incident photons for the targeted wavelength range. Researchers in [60] have used a SnO2/TiO2 materials heterojunction photo-anode with quantum dots design. The study highlighted that with quantum dots-based heterostructure in place, enhanced electron transfer characteristics would be achieved as compared to the bare TiO2 photo-electrode.

In one of the recent studies, Alvi et al. stated to achieve highest PCE for single junction $In_xGa_{1-x}N$/GaN photo-electrode-based PEC system [7]. The photo-electrode design was accomplished with the combination of InN quantum dots on $In_xGa_{1-x}N$ nanowire. The output observed was claimed to be 2.5 times better for PCE and hydrogen generation as compared to the case with only $In_xGa_{1-x}N$ nanowires photo-electrode and the actual output achieved for this unassisted water splitting was 9.3%. The superior output was credited to the nanostructured $In_xGa_{1-x}N$ surface morphology optimization and enhanced electronic properties with quantum dots. On the adverse side, the design and growth process still is a complex issue and required to be simplified with further efforts.

The discussion has given a glimpse of the larger perspective of the research and designs that are achieved for the photo-electrodes LMS for PEC systems. The gradual improvement in the design and growth process has lifted the PCE for the

PEC systems from 0.1% to just under 10% for single junction and just under 20% for complex tandem photo-electrode designs. However, there is still a lot is required to be achieved for the wider acceptance of the PEC by taking their CE ratio to the suitable limits. Main challenge in this regard is posed by the fabrication and growth processes and if, one can solve the design issues, the PEC has an enormous potential to solve the sustainable electrical energy generation and storage issues.

4. Present challenges and future perspective

An optimized PNS-based LMS for OSEH applications has established the fact through several researches that they have massive potential to overcome the inherently restricted absorption of the incident solar spectrum photons in the limited thickness of the active layer to generate useful current. It has been shown that with these PNS-based LMS, it is even possible to surpass fundamental limits now [51], as discussed. Although, the study has shown for SC applications that the overall results for PCE is beyond limiting values, the beyond limit absorption in the particular case was observed for only higher wavelength photons, whereas for lower wavelength region, it was still under the Lambertian values and thus there is still room available for further improvement. The further improvement can be achieved with reducing reflection losses and losses due to recombination within the structure over the largest possible incident wavelength range and ultimately enhancing the coupling of incident light in the entire wavelength range, to the active layer, to convert them into useful absorption.

But, the main restraint with these PNS-based devices is posed by the growth and fabrication schemes. Repeatedly designing such a structure requires sophisticated infrastructure that is very capital intensive, which is the basic hurdle in achieving the low CE ratio with these devices. There are basically two approaches are being used for the realization of PNS, namely, top-down and bottom-up approaches.

Most of the PNS are designed with top down approaches that include nanofabrication tools such as electron beam lithography, focused ion beam lithography, dip-pen lithography etc. These top-down lithography processes commonly used for designing PNS have the ability to achieve high precision in nm range. But these processes are usually very time consuming and it is really difficult to adapt the process for large area commercial applications. Also, although the processes are precise, they more or less are very time consuming, expensive, mostly machine dependent, having low throughput and it is really difficult to adapt the process for large area commercial applications. These factors put the major restriction in the commercialization of the PNS-based OSEH designs.

On the other hand, some of the nanostructures such as nanowires are mostly grown through so called bottom-up approaches. The process has a major advantage that with it one can achieve bulk production and high throughput at low cost. Various tools such as chemical vapor deposition (CVD), metalorganic chemical vapor deposition (MOCVD), molecular beam epitaxy (MBE), etc. are basically employed for these purposes [84] and growth is normally done in the presence of foreign catalysts. Although, the method is cost-effective and can provide bulk production at no time, it lacks precision and the process itself is mostly complicated. It is difficult to repeat the processes with same precision and many a times, even nonuniform and random profiles are achieved, even when one intends to generate periodic design. It is difficult to achieve the uniformity of the structure throughout the surface. The complex designs are really difficult to achieve with bottom-up approaches in the repeated manner.

As both top-down and bottom-up approaches has some limitations, one has to look toward a process that can have advantages of both. In this regard, during the recent years, several new techniques have rapidly came up to the arena due to their unique advantages including nanoimprint lithography [49], nanosphere lithography [85, 86], and self-assembled nano-masking scheme [87]. The processes promise to have all the major advantages related to top-down (due to the action of patterning the structured layers such as in a conventional lithography technique) and bottom up (due to the self-organization of the colloidal spheres) approaches and thus, are considered as viable inexpensive fabrication tools for producing regular and homogenous arrays of nanostructures with different sizes and need to be explored further.

An exciting scheme for OSEH for electrical energy generation that also deserves a special mention and seems very promising is the coupling of PEC with the conventional SC to electrolysis systems that can certainly improve the overall system's performance [21, 88]. The system can work for all day long operation and can be utilized to get electrical energy output during non-sunshine hours through electrochemical cells. Such a system can effectively developed as an alternate for the commonly used conventional SC and battery system and that too with better CE ratio, if designed with high PCE-based devices. The PCE of such a coupled system has been steadily improving [89, 90], and the indoor efficiency has already achieved the 30% mark [91]. However, the outdoor PCE is still limited, which is the foremost requirement for practical and commercial operations. Recently, Ota et al. [92] observed the record full day incident spectrum to the hydrogen conversion PCE of 18.78% for 470-W system under outdoor operation. It has been observed that to make this coupled system ready for the commercial and practical operations, it is primarily required to increase the involved individual components efficiency. The system, if optimized carefully, has a realistic potential to be considered for designing a reliable stand-alone system for future remote area lighting process as the overall set up will not be very heavy, and thus, the research efforts are required to be further in the area.

5. Conclusion

The chapter has discussed in details the technical perspective, advancements, requirement, and future trends for the PNS for designing the LMS for the solar energy harvesting applications that include solar cells and photoelectrochemical cells. The chapter put emphasis on the physical insight of the several engineered PNS designs that are being used for the purpose of LMS and how one can optimize the PNS for designing LMS. The chapter also discussed various bottlenecks that are still restricting the performance of the solar-to-electrical conversion process and highlighted several exciting methods for the further improvement in the performance of the device designs, which include utilizing tandem structures, improving the incident photons absorption through efficient spectrum harnessing using novel PNS, surface passivation processes, etc. From the discussion presented, one can appreciate the contribution that PNS-based LMS has made in the advancement of the SC and PEC devices and realize that it is due to the realistic potential of the PNS-based LMS that they have come down a long way, from merely a possibility to reality, to play a key role in achieving low CE systems for solar energy harvesting for sustainable development.

Conflict of interest

The author declares that there is no conflict of interest—whatsoever—related to financial or nonfinancial matters related to the work presented in the chapter.

Author details

Nikhil Deep Gupta
Centre for VLSI and Nanotechnology, Visvesvaraya National Institute of Technology, Nagpur, India

*Address all correspondence to: nikhildeepgupta@cvn.vnit.ac.in

References

[1] Brown K. United Nations Foundations [Online]. 2020. Available from: https://unfoundation.org/blog/post/5-global-issues-to-watch-in-2020/

[2] Ministry of New and Renewable Energy. Jawaharlal Nehru National Solar Mission: Towards Building SOLAR INDIA. New Delhi, India: Government of India; 2008

[3] NCPRE. Why Solar for Beginners [Online]. 2013. Available from: http://www.ncpre.iitb.ac.in

[4] Nelson J. The Physics of Solar Cell London. London, UK: Imperial College Press; 2008

[5] Green MA, Dunlop ED, Hohl-Ebinger J, Yoshita M, Kopidakis N, HA W. Solar cell efficiency tables (version 55). Progress in Photovoltaics. 2020;**28**(1):3-15

[6] Shockley W, Queisser HJ. Detailed balance limit of p-n junction solar cells. Journal of Applied Physics. 1961;**32**(510):510

[7] Alvi NH, Rodriguez PEDS, Hassan W, Zhou G, Willander M, Notzel R. Unassisted water splitting with 9.3% efficiency by a single quantum nanostructure photoelectrode. International Journal of Hydrogen Energy. 2019;**44**:19650-19657

[8] Naqvi A, Haug FJ, Soderstrom K, Battaglia C, Paeder V, Scharf T, et al. Angular behavior of the absorption limit in thin film silicon solar cells. Progress in Photovoltaics: Research and Applications. 2014;**22**(11):1147-1158

[9] Gupta ND, Janyani V. Design and analysis of light trapping in thin film GaAs solar cells using 2-D photonic crystal structures at front surface. IEEE Journal of Quantum Electronics. 2017;**53**(2):4800109

[10] Schuster CS, Bozzola A, Andreani LC, Krauss TF. How to assess light trapping structures versus a Lambertian Scatterer for solar cells? Optics Express. 2014;**22**(S2):A542-A551

[11] Green MA. Lambertian light trapping in textured solar cells and light-emitting diodes: Analytical solutions. Progress in Photovoltaics: Research and Applications. 2002;**10**:235-241

[12] Feng NN, Zhou GR, Huang WP. Space mapping technique for design optimization of antireflection coatings for photonic devices. Journal of Lightwave Technology. 2003;**21**(1):281-285

[13] Gupta ND, Janyani V. Lambertian and photonic light trapping analysis with thickness for GaAs solar cells based on 2D periodic pattern. IET Optoelectronics. 2017;**11**(5):217-224

[14] Zhou D, Biswas R. Photonic crystals enhanced light trapping in thin film solar cells. Journal of Applied Physics. 2008;**103**:093102

[15] Gupta ND, Janyani V. Design and optimization of photonic crystal diffraction grating based efficient light trapping structure for GaAs thin film solar cell. Journal of Nanoelectronics and Optoelectronics. 2016;**11**(4):407-415

[16] Davydov VY, Klochikhin AA, Seisyan RP, Emtsev VV, Ivanov SV, Bechstedt F, et al. Absorption and emission of hexagonal InN evidence of narrow fundamental band gap. Physica Status Solidi B: Basic Solid State Physics. 2002;**229**(3):R1-R3

[17] Wu J, Walukiewicz W, Yu KM, Ager JW III, Haller EE, Lu H, et al. Unusual properties of the fundamental band gap of InN. Applied Physics Letters. 2002;**80**(21):3967-3969

[18] Gupta ND, Janyani V, Mathew M, Maun M, Singh R. Design and fabrication of InGaN/GaN Superlattice based solar cell using photonic crystal structure at the front surface. Journal of Nanophotonics. 2018;**12**(4):043505

[19] Benton J, Bai J, Wang T. Enhancement in solar hydrogen generation efficiency using a GaN-based nanorod structure. Applied Physics Letters. 2013;**102**:173905

[20] Cai XM, Zeng SW, Zhang BP. Fabrication and characterization of InGaN p-i-n homojunction solar cell. Applied Physics Letters. 2009;**95**

[21] Sugiyama M, Fujii K, Nakamura S. Solar to Chemical Energy Conversion—Theory and Application. Tokyo: Springer; 2016

[22] Lobanova AV, Kolesnikova AL, Romanov AE, Karpov SY, Rudinsky ME, Yakovlev EV. Mechanism of stress relaxation in (0001) InGaN/GaN via formation of V-shaped dislocation half-loops. Applied Physics Letters. 2013;**103**(15):152106

[23] Nakamura S, Pearton S, Fasol G. The Blue Laser Diode. 2nd ed. Berlin: Springer-Verlag; 2000

[24] Yan B, Owens JM, Jiang C, Guha S. High-efficiency amorphous silicon alloy based solar cells and modules. MRS Symp Proc. 2005;**A23.3**:862

[25] Yablonovitch E, Cody GD. Intensity enhancement in textured optical sheets for solar cells. IEEE Transactions on Electron Devices. 1982;**29**(2):300-305

[26] Springer J, Poruba A, Mullerova L, Vanecek M, Rech O, K B. Absorption loss at nanorough silver back reflector of thin-film silicon solar cells. Journal of Applied Physics. 2004;**95**:1427-1429

[27] Zhao S, Nguyen HPT, Kibria MG, Mi Z. III-nitride nanowire optoelectronics. Progress in Quantum Electronics. 2015;**44**:14-68

[28] Gupta ND, Janyani V, Mathew M. Light trapping in p-i-n superlattice based InGaN/GaN solar cells using photonic crystals. Optical and Quantum Electronics. 2016;**48**(11):502, 1-517

[29] Gupta ND, Janyani V. Dense wavelength division demultiplexing using photonic crystal waveguides based on cavity resonance. Optik. 2014;**125**(19):5833-5836

[30] Paliwal A, Singh K, Mathew M. Effects of an undoped-InGaN waveguide on the optical confinement and carrier dynamics of InGaN laser diodes. Laser Physics. 2018;**28**(12):126204

[31] Leung SF, Zhang Q, Xiu F, Yu D, Ho JC, Li D, et al. Light management with nanostructures for optoelectronic devices. Journal of Physical Chemistry Letters. 2014;**5**(8):1479-1495

[32] Osterloh FE. Inorganic nanostructures for photoelectrochemical and photocatalytic water splitting. Chemical Society Reviews. 2013;**42**:2294-2320

[33] Chen X, Li C, Graetzel M, Kostecki R, Mao SS. Nanomaterials for renewable energy production and storage. Chemical Society Reviews. 2012;**41**:7909-7937

[34] Bermel P, Luo C, Zeng L, Kimerling LC, Joannopoulos JD. Improving thin-film crystalline silicon solar cell efficiencies with photonic crystals. Optics Express. 2007 Dec;**15**(25):16986-17000

[35] Feng N-N, Michel J, Zeng L, Liu J, Hong C-Y, Kimerling LC, et al. Design of highly efficient light-trapping structures for thin-film crystalline silicon solar cells. IEEE Transactions on Electron Devices. 2007;**54**(8):1926-1933

[36] Bozzola A, Liscidini M, Andreani LC. Photonic light-trapping versus Lambertian limits in thin film silicon solar cells with 1D and 2D periodic patterns. Optics Express. 2012;**20**(S2):224-243

[37] Zanotto S, Liscidini M, Andreani LC. Light trapping regimes in thin-film silicon solar cells with a photonic pattern. Optics Express. 2010;**18**(5):4260-4274

[38] Joannopoulos DJ. Photonic Crystals—Molding the Flow of Light. New Jersey: Princeton University Press; 1995

[39] Eyderman S, John S. Light-trapping and recycling for extraordinary power conversion in ultra-thin gallium-arsenide solar cells. Scientific Reports. June 2016;**6**(28303):1-7

[40] Ding H, Lalouat L, Gonzalez-Acevedo B, Orobtchouk R, Seassal C, Drouard E. Design rules for net absorption enhancement in pseudo-disordered photonic crystal for thin film solar cells. Optics Express. 2016;**24**(6):A650-A666

[41] Sheng X, Broderick LZ, Kimerling LC. Photonic crystal structures for light trapping in thin-film Si solar cells: Modeling, process and optimizations. Optics Communication. 2014;**314**:41-47

[42] Wang X, Khan MR, Gray JL, Alam MA, Lundstrom MS. Design of GaAs solar cells operating close to the Shockley–Queisser limit. IEEE Journal of Photovoltaics. 2013;**3**(2):737-744

[43] Chutinan A, Kherani NP, Zukotynski S. High-efficiency photonic crystal solar cell architecture. Optics Express. 2009;**17**(11):8871-8878

[44] Gupta ND. Absorption enhancement in hole Interface layer free perovskite solar cells using periodic photonic nanostructures. Optics and Laser Technology. 2019;**115**:20-31

[45] Munday JN. The effect of photonic bandgap materials on the Shockley-Queisser limit. Journal of Applied Physics. 2012;**112**:064501

[46] Demesy G, John S. Solar energy trapping with modulated silicon nanowire photonic crystals. Journal of Applied Physics. 2012;**112**:074326

[47] Mallick SB, Agrawal M, Peumans P. Optimal light trapping in ultra-thin photonic crystal crystalline silicon solar cells. Optics Express. 2010;**18**(6):5691

[48] Eyderman S, Deinega A, John S. Near perfect solar absorption in ultra-thin-film GaAs photonic crystals. Journal of Materials Chemistry A. 2014;**2**(3):761-769

[49] Peer A, Biswas R, Park JM, Shinar R, Shinar J. Light management in perovskite solar cells and organic LEDs with microlens arrays. Optics Express. 2017;**25**(9):10704

[50] SaJS B. Photonic crystal light trapping: Beyond 30% conversion efficiency for silicon photovoltaics. APL Photonics. 2020;**5**:020902

[51] Hsieh ML, Kaiser A, Bhattacharya S, John S, Lin SY. Experimental demonstration of broadband solar absorption beyond the Lambertian limit in certain thin silicon photonic crystals. Scientific Reports. 2020;**10**:11857

[52] Bielawny A, Rockstuhl C, Lederer F, Wehrspohn RB. Intermediate reflectors for enhanced top cell performance in photovoltaic thin-film tandem cells. Optics Express. 2009;**17**(10):8439-8446

[53] Mutitu JG, Shi S, Chen C, Creazzo T, Barnett A, Honsber C, et al. Thin film silicon solar cell design based on photonic crystal and diffractive

grating structures. Optics Express. 2008;**16**(19):15238

[54] Green MA. Commercial progress and challenges for photovoltaics. Nature Energy. 2016;**1**:15015

[55] Hossain MI, Qarony W, Ma S, Zeng L, Knipp D, Tsang Y. Perovskite/silicon tandem solar cells: From detailed balance limit calculations to photon management. Nano-Micro Letters. 2019;**11**:58

[56] Wang K, Jin Z, Liang L, Bian H, Bai D, Wang H, et al. All inorganic cesium lead iodide perovskite solar cells with stabilized efficiency beyond 15%. Nature Communications. 2018;**9**:4544

[57] Werner J, Nogay G, Sahli F, TCJ Y, Brauninger M, et al. Complex refractive indices of cesium–formamidinium-based mixed-halide perovskites with optical band gaps from 1.5 to 1.8 eV. ACS Energy Letters. 2018;**3**:742-747

[58] Steiner MA, Barraugh CD, Aldridge CW, Alvarez IB, Friedman DJ, Ekins-Daukes NJ, et al. Photoelectrochemical water splitting using strain-balanced multiple quantum well photovoltaic cells. Sustainable Energy & Fuels. 2019;**3**:2837-2844

[59] Rodrigueza I, Atienzar P, Ramiro-Manzano F, Meseguer F, Corma A, Garcia H. Photonic crystals for applications in photoelectrochemical processes: Photoelectrochemical solar cells with inverse opal topology. Photonics and Nanostructures. 2005;**2**(2-3):148-154

[60] Basu K, Zhang H, Zhao H, Bhattacharya S, Navarro-Pardo F, Datta PK, et al. Highly stable photoelectrochemical cells for hydrogen production using a SnO2–TiO2/quantum dot heterostructured photoanode. Nanoscale. 2018;**10**:15273-15284

[61] Dolmanan SB, Lai SC, Ke L, Loh WW, Jiao ZH, Sun XW. Improved photoelectrochemical cell with carbon nanotubes. IEEE Electron Device Letters. 2010;**31**(7):734-736

[62] Collazo R, Dietz N. The group III-nitride material class: From preparation to perspectives in photoelectrocatalysis. In: Photoelectrochemical Water Splitting: Issues and Perspectives. Cambridge, UK: RSC Publishing; 2013. pp. 193-222

[63] Wang D, Pierre A, Kibria MG, Cui K, Han X, Bevan KH, et al. Wafer-level Photocatalytic water splitting on GaN nanowire arrays grown by molecular beam Epitaxy. Nano Letters. 2011;**11**(6):2353-2357

[64] Maeda K, Teramura K, Saito N, Inoue Y, Domen K. Photocatalytic overall water splitting on gallium nitride powder. Bulletin of the Chemical Society of Japan. 2007;**80**(5):1004

[65] Kida TY, Minami Y, Guan G, Nagano M, Akiyama M, Yoshida A. Photocatalytic activity of gallium nitride for producing hydrogen from water under light irradiation. Journal of Materials Science. 2006;**41**:3527-3534

[66] Wang J, Pedroza LS, Poissier A, Fernández-Serra MV. Water dissociation at the GaN$(10\bar{1}0)$ surface: Structure, dynamics and surface acidity. Journal or Physical Chemistry C. 2012;**116**(27):14382-14389

[67] Jung HS, Hong YJ, Li Y, Cho J, Kim YJ, Yi GC. Photocatalysis using GaN nanowires. ACS Nano. 2008;**2**(4):637-642

[68] Zhang Z, Yates JT. Band bending in semiconductors: Chemical and physical consequences at surfaces and interfaces. Chemical Reviews. 2012;**112**(10):5520-5551

[69] Kibria MG, Chowdhury FA, Zhao S, Trudeau ML, Guo H, Mi Z. Defect-engineered GaN:Mg nanowire arrays for overall water splitting under

violet light. Applied Physics Letters. 2015;**106**:113105

[70] Wu J. When group-III nitrides go infrared: New properties and perspectives. Journal of Applied Physics. 2009;**106**:011101

[71] Benton J, Bai J, Wang T. Utilisation of GaN and InGaN/GaN with nanoporous structures for water splitting. Applied Physics Letters. 2014;**105**(22):2012-2017

[72] Kibria MG, Nguyen HPT, Cui K, Zhao S, Liu D, Guo H, et al. One-step overall water splitting under visible light using multiband InGaN/GaN nanowire heterostructures. ACS Nano. 2013;**7**(9):7886-7893

[73] Kibria M, Chowdhury F, Trudeau M, Guo H, Mi Z. Dye-sensitized InGaN nanowire arrays for efficient hydrogen production under visible light irradiation. Nanotechnology. 2015;**26**(28):285401

[74] Kibria MG, Zhao S, Chowdhury F, Wang Q, Nguyen HPT, Trudeau ML, et al. Tuning the surface Fermi level on p-type gallium nitride nanowires for. Nature Communications. 2014;**5**:1-6

[75] Chowdhury FA, Mi Z, Kibria MG, Trudeau ML. Group III-nitride nanowire structures for photocatalytic hydrogen evolution under visible light irradiation. Applied Physics Letters. 2015;**3**:104408

[76] Alvi NH, Soto Rodriguez PED, Kumar P, Gómez VJ, Aseev P, Alvi AH, et al. Photoelectrochemical water splitting and hydrogen generation by a spontaneously formed InGaN nanowall network. Applied Physics Letters. 2014;**104**:223104

[77] Caccamo L, Hartmann J, Fàbrega C, Estradé S, Lilienkamp G, Prades JD, et al. Band engineered epitaxial 3D GaN-InGaN core-shell rod arrays as an advanced photoanode for visible-light-driven water splitting. ACS Applied Materials & Interfaces. 2014;**6**(4):2235-2240

[78] Ebaid M, Kang JH, Lim SH, Ha JS, Lee JK, Cho YH, et al. Enhanced solar hydrogen generation of high density, high aspect ratio, coaxial InGaN/GaN multi-quantum well nanowires. Nano Energy. 2015;**12**:215-223

[79] Hwang YJ, Wu CH, Hahn C, Jeong HE, Yang P. Si/InGaN Core/shell hierarchical nanowire arrays and their photoelectrochemical properties. Nano Letters. 2012;**12**:1678-1682

[80] Young JL, Steiner MA, Döscher H, France RM, Turner JA, Deutsch TG. Direct solar-to-hydrogen conversion via inverted metamorphic multi-junction semiconductor architectures. Nature Energy. 2017;**2**:17028

[81] Chowdhury FA, Trudeau ML, Guo H, Mi Z. A photochemical diode artificial photosynthesis system for unassisted high efficiency overall pure water splitting. Nature Communications. 2018;**9**:1707

[82] Kharche N, Hybertsen MS, Muckerman JT. Computational investigation of structural and electronic properties of aqueous interfaces of GaN, ZnO, and a GaN/ZnO alloy. Physical Chemistry Chemical Physics. 2014;**16**:12057-12066

[83] Kibria MG, Qiao R, Yang W, Boukahil I, Kong X, Chowdhury FA, et al. Atomic-scale origin of long-term stability and high performance of p-GaN nanowire arrays for photocatalytic overall pure water splitting. Advanced Materials. 2016;**28**:8388-8397

[84] Zhao S, Le BH, Liu DP, Liu XD, Kibria MG, Szkopek T, et al. p-type InN nanowires. ACS Nano Letters. 2013;**13**(11):5509-5513

[85] Wu LY, Ross BM, Lee LP. Optical properties of the crescent-shaped

nanohole antenna. Nano Letters. 2009;**9**(5):1956-1961

[86] Zhang Q, Ghosh S, Samitsu S, Peng X, Ichinose I. Ultrathin freestanding nanoporous membranes prepared from polystyrene nanoparticles. Journal of Materials Chemistry. 2011;**21**(6):1684-1688

[87] Mathew M, Sodabanalu H, Sugiyama M, Nakano Y. Orange/yellow InGaN/AlN nanodisk light emitting diodes. Physica Status Solidi C: Current Topics in Solid State Physics. 2013;(11):1525

[88] Chu S, Li W, Yan Y, Hamann T, Shih I, Wang D, et al. Roadmap on solar water splitting: Current status and future prospects. Nano Futures. 2017;**1**: 022001:1-29

[89] Barbir F. PEM electrolysis for production of hydrogen from renewable energy sources. Solar Energy. 2005;**78**(5):661-669

[90] Ager JW, Shaner MR, Walczak KA, Sharp ID, Ardo S. Experimental demonstrations of spontaneous, solar-driven photoelectrochemical water splitting. Energy & Environmental Science. 2015;**8**:2811-2824

[91] Jia J, Seitz LC, Benck JD, Huo Y, Chen Y, Ng JW, et al. Solar water splitting by photovoltaic-electrolysis with a solar-to-hydrogen efficiency over 30. Nature Communications. 2016;7:13237

[92] Ota Y, Yamashita D, Nakao H, Yonezawa Y, Nakashima Y, Ebe H, et al. Highly efficient 470 W solar-to-hydrogen conversion system based on concentrator photovoltaic modules with dynamic control of operating point. Applied Physics Express. 2018;**11**:077101

Chapter 6

Single Crystal Hybrid Perovskite Optoelectronics: Progress and Perspectives

Feng Li

Abstract

Organic–inorganic hybrid perovskites, which combine the superior optical and electronic properties and solution-processed manufacturing, have emerged as a new class of revolutionary optoelectronic devices with the potential for various practical applications. Encouraged by the advantages of longer carrier diffusion length, higher carrier mobility and lower trap densities as compared to the polycrystalline counterparts', increasing research attention has focused on preparation and optimization of perovskite crystal candidates, via using various facile growth techniques, for the development of a wide range of optoelectronic applications. This chapter presents a comprehensive review of recent advances in the field of optoelectronic technologies based on different forms of single crystals, including bulk crystals and thin ones, with emphasis placed on the optimization of crystals and the relationship among the charge-carrier transport, operation mechanism, device architecture, and device performance. First, we introduce the main methods used to prepare bulk and thin single crystals, and analyze several aspects of their properties. Thereafter, the applications of single crystals into solar cells, photodetectors, light-emitting diodes, and lasers, are discussed in depth. Finally, we summarize the challenges of perovskite single crystals and propose further improvements in the synthesis approaches and device applications.

Keywords: hybrid perovskite crystals, optoelectronics, solar cells, photodetectors, LEDs

1. Introduction

Perovskite-structured materials have received increasing attention, since being discovered in the 1830s, because of their rich physical properties [1]. As shown in **Figure 1a** [2], the general chemical formula for such compounds is ABX_3, in which A and B are different cations, and X is an anion that bonds to both the A and B cations. Owing to the flexibility of bond angles inherent in the perovskite structure, there are many different distortions that can occur from the ideal structure. Importantly, A can be organic cations, like methylammonium (MA^+) or formamidinium (FA^+) [4–8], B can be metal ions, such as Pb^{2+} and Sn^{2+} [9–12], and X is usually halide ions [13], and such a class of materials is known

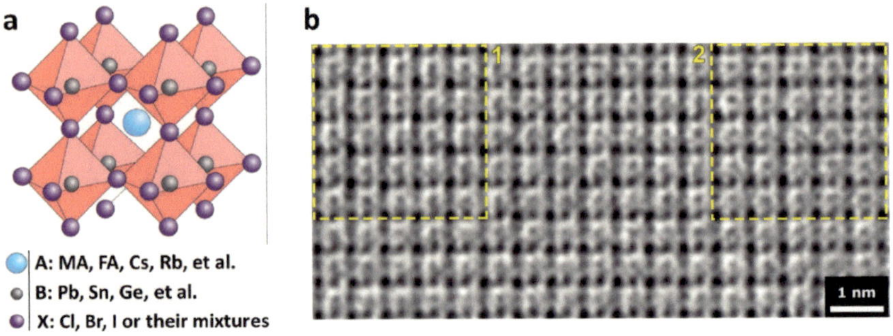

Figure 1.
1a, perovskite crystal structure. Nature Photonics *[2], copyright 2014.* ***2b****, CTF-corrected high-revolution TEM image.* Science *[3], copyright 2018.*

as organic–inorganic hybrid perovskites. It was reported that a stable structure of hybrid perovskites can form where 0.81 < *T.F.* (tolerance factor) < 1.1 and 0.44 < *O.F.* (octahedral factor) < 0.90 [14]. X-ray diffraction (XRD) measurements were widely used to characterize their structures. As for $MAPbBr_3$ and $MAPbI_3$ crystals, XRD measurements displayed the excellent single crystal properties [15]. Transmission electron microscopy (TEM) measurements were performed to provide a more intuitive picture of perovskite crystals structures (**Figure 1b**), via using contrast-transfer-function corrected method to overcome their electron beam-sensitive property [3]. After the first attempt to employ hybrid perovskite films as active sensitizers into photovoltaic devices [16], hybrid perovskite solar cells have continued to set new efficiency benchmarks [17–23], due to the excellent properties, such as ease of processing, tunable optical band gaps [24, 25], long carrier diffusion length [26], and low trap density [15], as well as large absorption coefficients and high photoluminescence (PL) efficiency [27, 28], and their relatively high power conversion efficiency (PCE) has been increased to as high as 25.2% [29]. Moreover, Leveraging their promising features, hybrid perovskites also have the potential for employment in other optoelectronic applications, including photodetectors [30], transistors [31], phototransistors [32], light-emitting diodes (LEDs) [33], and lasers [34].

However, a vast array of prior research on perovskite optoelectronic devices has been centered on polycrystalline films. The polycrystalline samples usually suffer from grain boundaries, relatively higher trap densities and defects, and low stability, which would obviously obscure their potential in applications [35–37]. More recently, researchers have paid more attention to perovskite single crystals, which possess promising characteristics of no grain boundaries [15], relatively low trap density [38], large charge carrier mobility, and long carrier diffusion length [39–41]. In this regard, extensive efforts are being devoted to developing effective methods to improve the perovskite crystal quality and optimize the device performance. Existing in the forms of bulk or thin crystals, perovskite crystal samples have been widely applied in various optoelectronic applications [39, 42], and have made rapid and great strides in research progress [43–46].

In this chapter, we aim to summarize the recent achievements, ongoing progress, and the challenges to date in the area of hybrid perovskite single crystals, practically MA-based ones ($MAPbX_3$, X = Cl, Br, and I), from the perspective of both materials and devices with an emphasis placed on the optimization of crystal quality, and provide an outlook on the opportunities offered by this emerging family of materials in field of optoelectronic applications.

2. Growth of hybrid perovskite single crystals

2.1 Bulk single crystals

2.1.1 Solution temperature-lowering (STL) method

According to the lower solubility of $MAPbX_3$ in HX (X = Cl, Br, and I) solution as the temperature decreases, Tao's group introduced the STL method to synthesize a $MAPbI_3$ bulk single crystal (**Figure 2a**) [47]. After the reaction between methylamine (CH_3NH_2) and hydro-iodic acid (HI) in a cold atmosphere, the obtained white microcrystal MAI was reacted with $Pb(CH_3COOH)_2{\bullet}3H_2O$ in aqueous HI, and the solution was then cooled to 40°C. A 10 mm × 10 mm × 8 mm black $MAPbI_3$ single crystal was grown in about one month (**Figure 2b**). Lin's group discovered a more efficient way, and they synthesized the single crystals with a size of 5 mm in just around 10 days [48]. Lin et al. selected high-quality seeds and dropped them back into fresh solution and obtained single crystals sized up to 1 cm (**Figure 2c**). Furthermore, $MAPbBr_{3-x}Cl_x$ and $MAPbI_{3-x}Br_x$ mixed-halide perovskite crystals were studied using such method [49]. Hydro-bromic acid with hydrochloric acid or hydro-iodic acid were mixed in different molar ratios into methylamine and lead (II) acetate solution to fabricate single-halide and mixed-halide perovskite crystals (**Figure 2d**). The time consuming factor is the biggest drawback of this method, which has indirectly led to the domination of other crystallization methods.

2.1.2 Inverse temperature crystallization (ITC) method

As a radically faster perovskite crystal synthesis approach, the ITC method has widely been applied in recent years. It was observed that the exhibited crystals from such method can be shape-controlled, higher quality, and obtained quicker compared with other growth techniques. Bakr et al. introduced this method to rapidly grow high-quality bulk crystals [50]. As shown in **Figure 2e**, an orange $MAPbBr_3$ crystal and a black $MAPbI_3$ crystal were grown within 3 hours. Chen's group further

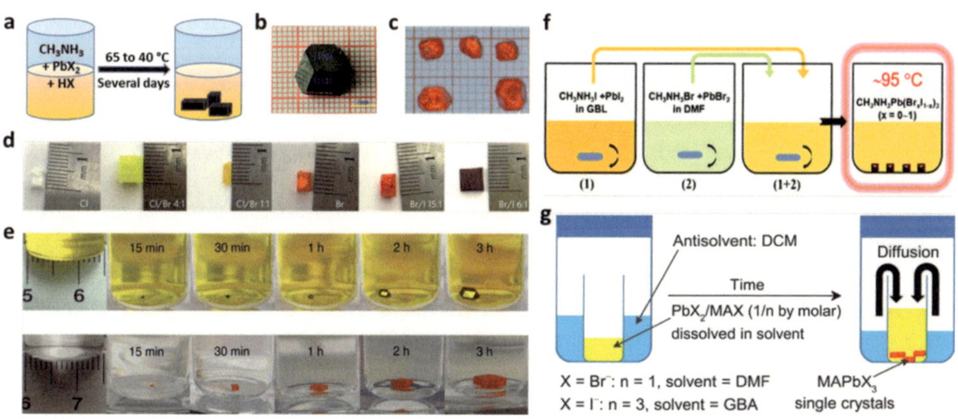

Figure 2.
***2a**, schematic of STL method. **2b**, image of $MAPbI_3$ with {100} and {112} facets.* CrystEngComm *[47], copyright 2015. **2c**, $MAPbBr_3$ crystals from STL method.* J. Cryst. Growth *[48], copyright 2015. **2d**, photographs of perovskite crystals with different halide ratio.* Nature Photonics *[49], copyright 2015. **2e**, $MAPbI_3$ and $MAPbBr_3$ crystals growth at different time intervals.* Nature Commun. *[50], Copyright 2015. **2f**, schematic of crystals growth.* J. Mater. Chem. C *[51], copyright 2016. **2 g**, schematic of AVC method.* Science *[15], copyright 2015.*

studied the effect of molar ratio of MAX and PbX_2 in the precursor solutions on the crystal quality [52], e.g., perovskite crystals with different sizes and shapes were obtained after a 6-hour ITC crystallization process when changing the MAX: PbX_2 ratios from 1:1 to 2:1.

With an aim of growing a large-sized bulk perovskite crystal, such ITC method was further modified. Using such technique, the strategy of incorporation of seed crystal growth has been proven to be favorable for single crystals as large as convenient. Liu's group reported various large-sized perovskite crystals via using the modified ITC method, from which a number of larger-sized crystal (7 mm) were obtained through choosing good-quality seed crystals and repeating and carefully controlling the ITC process several times (**Figure 2f**). Moreover, Liu's group also successfully grew $MAPb(Br_xI_{1-x})_3$ single crystals with a finely-tuned bandgap [51]. The application of the different solubility of different perovskite single crystals at varying temperatures contributes to the time-saving feature of such ITC method.

2.1.3 Anti-solvent vapor-assisted crystallization (AVC) method

Another main method to grow perovskite crystals is the AVC method (**Figure 2g**), which was first introduced from Bakr's group [15]. In this method, the solvent plays a significant role because two or more solvents should be selected, of which one should be a good solvent that is less volatile, and the other is a bad solvent that is more volatile. The principle of this method can be described as follows: when the bad solvent slowly diffuses into the precursor solution, the proficiency of the crystal formation increases at the bottom of the sample vial owing to the insolubility of the material in the bad solvent. Other groups, like Loi's group and Cao's group, also applied this method to obtain the high-quality crystals [38, 53]. Although the AVC method costs more time than the ITC method, its temperature-irrelevant characteristic is appealing to its widespread use.

2.2 Thin single crystals

Bulk perovskite single crystals with thick sizes may cause the increase of charge recombination, which would lead to the degradation of their device performance and impede the practical applications. In this regard, growing thin perovskite crystals with a large area represents an effective approach to overcome the above obstacle and thus advances the further practical applications. Bakr et al. introduced a cavitation-triggered asymmetrical crystallization strategy, in which a very short ultrasonic pulse ($\approx$1 s) was applied in the solution to reach a low supersaturation level with anti-solvent vapor diffusion and a thin crystal with several-micrometers grew on the substrates within hours (**Figure 3a**) [54]. Liu's group synthesized perovskite crystal wafers with a much thinner thickness using a dynamic flow micro-reactor system [55]. They put two thin glass slides in parallel into a container with a predefined separation to grow single crystals within the slit channel, as shown in **Figure 3b**. Su's group further used a space-limited ITC method and grew a 120-cm^2 single crystal on fluorine-doped tin oxide (FTO)-coated glass, of which the operation and the obtained 0.4-mm-thin single crystal are shown in **Figure 3c** [56]. Meanwhile, Wan et al. reported a space-confined solution-processed method to grow the perovskite single-crystalline films with adjustable thickness from nanometers to micrometers (**Figure 3d**) [57]. Benefitting from the capillary pressure, the perovskite precursor solution filled the whole space between two clean flat substrates, which were clipped together and dipped in the solution.

Currently, more promising approaches have been employed to grow thin single crystals with high quality and large scale. A one-step printing geometrically-confined

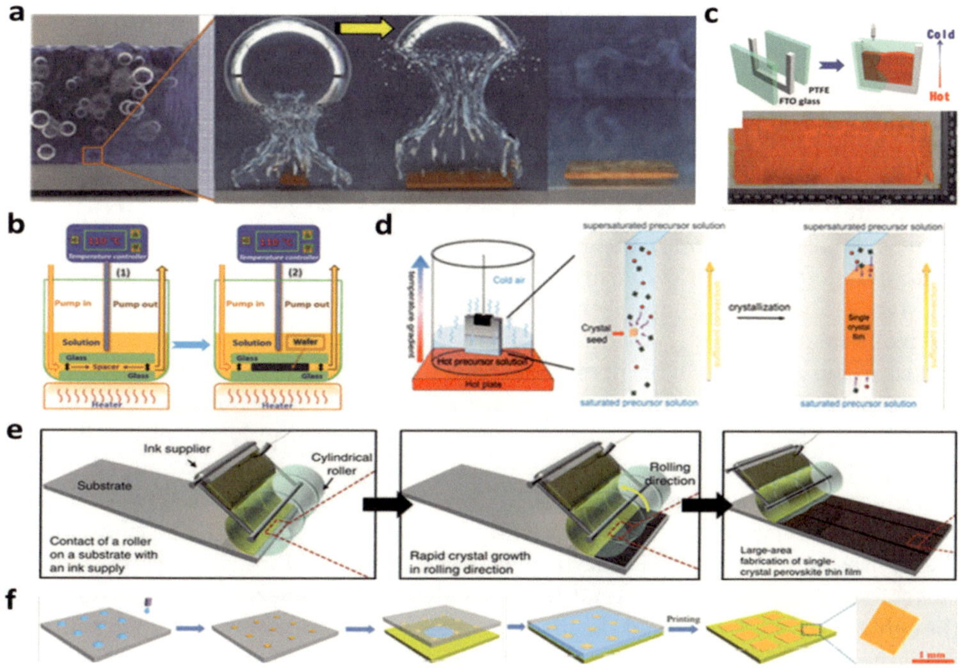

Figure 3.
***3a,** schematic of cavitation-triggered asymmetrical method.* Adv. Mater. *[54], Copyright 2016.* ***3b,** schematic of ultrathin crystal wafer growth.* Adv. Mater. *[55], Copyright 2016.* ***3c,** schematic of the laminar $MAPbBr_3$ crystal films preparation.* Adv. Mater. *[56], Copyright 2017.* ***3d,** schematic for the growth of perovskite thin crystals.* J. Am. Chem. Soc. *[57], copyright 2016.* ***3e,** schematic of geometrically-confined lateral crystal growth method.* Nature Commun. *[58], Copyright 2017.* ***3f,** schematic of the scalable growth for perovskite crystal films using an inkjet printing method.* Sci. Adv. *[59], Copyright 2018.*

lateral crystal growth method (**Figure 3e**) was introduced by Sung's group to obtain a large-scaled single crystal [58]. During the process, a cylindrical metal roller with a flexible poly-(dimethyl-siloxane) (PDMS) mold was wrapped and then rolled on a preheated SiO_2 substrate (180°C) with an ink supplier filled with the precursor solution. Alternatively, millimeter-sized single crystals were synthesized by Song's group by a facile seed-inkjet-printing approach (**Figure 3f**) [59]. Perovskite precursor solution was injected onto a silicon wafer, and then the ordered seeds were formed on the substrate with the evaporation of the droplets. Thereafter, the substrate with a saturated perovskite solution was covered and the single crystals can be grew as the solvent dried at room temperature. Seeds were used to inhibit the random nucleation and trigger the growth of single crystals.

As discussed above, some optimized space-limited approaches have been introduced and developed to synthesize perovskite thin crystals in recent years. Especially, size–/thickness-controlled thin crystals have also been widely used in various optoelectronic devices. With the aim to growing large-scaled and thickness-controlled thin crystals with longer carrier diffusion lengths, fewer defects, and higher efficiency, more promising strategies will be rewarding in the future.

3. Optoelectronic characterizations of perovskite single crystals

3.1 Optical properties

There are two normal ways to study the optical properties of hybrid perovskite crystals: absorption and PL measurements. Bakr et al. characterized the steady-state absorption and PL properties for $MAPbBr_3$ and $MAPbI_3$ crystals, as shown in **Figure 4a** and **b** [50]. Sharp band edges were observed in the absorption plots and the band gap values were determined to be 2.18 eV for $MAPbBr_3$ crystals and 1.51 eV for $MAPbI_3$ crystals; while the PL intensity peaks are located at 574 nm for $MAPbBr_3$ and 820 nm for $MAPbI_3$. As for the $MAPbCl_3$ one, absorption measurement result revealed an edge at 435 nm (**Figure 4c**) [60]. Clearly, the optical absorption of perovskite crystals exhibited a clear-cut sharp band edge, which indicated that the single crystals are predominantly free from grain boundaries and have relatively low structural defects and trap densities.

More recently, there have been more broad publications on the apparent disparity in optical properties (i.e., absorption and PL) between perovskite single crystals and thin films, which can be attributed to the incorrect measurements as a result of reabsorption effects. Snaith's group performed a detailed investigation of the optical properties of $MAPbBr_3$ crystals as compared to those of the polycrystalline films by employing light transmission spectroscopy, ellipsometry, and spatially resolved and time-resolved PL spectroscopy [61]. They showed that the optical properties of the perovskite crystals were almost identical to those of polycrystalline films, and their observations indicated that the perovskite polycrystalline films were much closer to possessing 'single-crystal-like' optoelectronic properties than previously thought, and also highlighted the discrepancies in the estimation of trap densities from the electronic and optical methods (**Figure 4d**). For the further development of perovskite crystals, more detailed experimental investigations combined with theoretical calculations that focus on the optical features are required, which would assist in the preparation of the high-quality perovskite single crystals and the development of the high-performance device applications.

3.2 Charge transport properties

As for hybrid perovskite crystals, in addition to the remarkable optical properties, their promising electrical properties have caught the great attention. In general, there are five common methods to measure the transport mobilities in perovskite crystals, including the space charge limited current (SCLC), time-of-flight (TOF), Hall Effect, THz pulse and field-effect transistor (FET) measurement methods. Among these methods, the SCLC method is widely employed to determine the carrier mobility and trap density of perovskite crystals. The current–voltage (*I-V*) curve can be divided into three parts: the first region, where an Ohmic contact exists, hence the conductivity can be estimated; the second region is the trap-filling region, which is increased sharply at trap-filled limit voltage (V_{TFL}); and the third region, known as the child region. Trap density (n_{trap}) can be obtained by following the relation: $n_{trap} = (2V_{TFL}\varepsilon\varepsilon_0)/(eL^2)$, where ε_0 is the vacuum permittivity, ε is the relative dielectric constant, L is the crystal thickness, and e is the electron

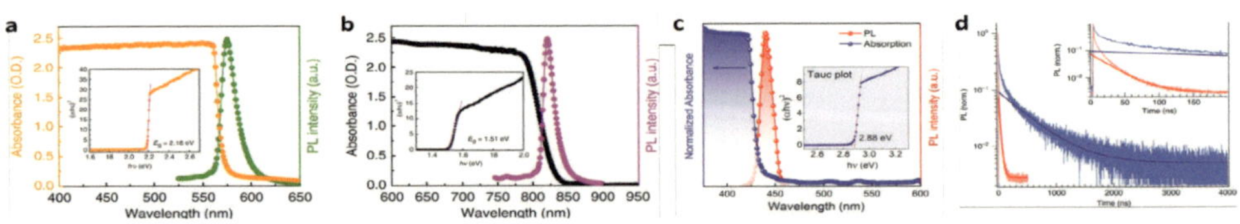

Figure 4.
*Steady-state absorption (**4a**) and PL spectra (**4b**) of $MAPbBr_3$ and $MAPbI_3$ crystals, respectively.* Nature Commun. *[50], Copyright 2015.* **4c**, *steady-state absorption and PL spectra of $MAPbCl_3$ crystal. Insets: Band gap for the above single crystals.* J. Phys. Chem. Lett. *[60], Copyright 2015,* **4d**, *normalized PL decays for $MAPbBr_3$ film (red) and crystal (blue) excited at 447 nm. Inset shows the zoom on the shorter time scale.* Nature Commun. *[61], Copyright 2017.*

charge. Moreover, the mobility (μ) is determined by fitting the *I-V* curve with Mott-Gurney's law: $\mu = (8JL^3)/(9\varepsilon\varepsilon_0 V^2)$, where J is the current density. Liu's group designed the hole-only device (**Figure 5a**), and a large hole mobility of 67.27 cm^2/Vs was estimated [62]. An SCLC method was also applied on $MAPbBr_3$ crystals, with an n_{trap} of 5.8×10^9 cm^{-3} and a μ of 38 cm^2/Vs [15]. *I-V* response of a $MAPbCl_3$ crystal was measured by Bakr's group with $n_{trap} = 3.1 \times 10^{10}$ cm^{-3} and $\mu = 42$ cm^2/Vs [60]. Another method to measure the μ is the TOF method. Bakr's group obtained the μ via using the TOF method (**Figure 5b**) [15], from which μ can be defined by the equation: $\mu = d^2/(V\tau_t)$, where d is the sample thickness, V is the applied voltage, and τ_t is the transit time that be provided by the transient current under different driving voltages [67, 68]. The same method was also applied by Huang's group and the electron μ was verified to be 24.0 ± 6.8 cm^2/Vs (**Figure 5c**) [63]. Apart from the above two methods, Bakr et al. also carried out the complementary Hall Effect measurements on perovskite crystals, confirming the μ ranging from 20 to 60 cm^2/Vs [15]. Meanwhile, Huang's group applies the Hall Effect measurement [68], and they showed the crystals were slightly p-doped with a low free holes concentration. Thereafter, Podzorov's group increased the conductivity of $MAPbBr_3$ single crystals by sputtering Ti on the flat-faceted single crystal to form Hall bars (**Figure 5d**) [64], from which the Hall mobility was calculated to be 10 cm^2/Vs.

Although the above measurement approaches have been widely used in the perovskite crystals, the obtained results from different groups are sometimes different. Sargent et al. demonstrated that one main challenge that may explain these order-of-magnitude discrepancies is that the Hall Effect, TOF, and SCLC methods all probe the mobilities near the respective Fermi levels during the experiments, and the (non-equilibrium, high-injection-level) Fermi level is widely different in each experiment [64]. In this regard, they developed a contactless method to measure the mobility of a perovskite crystal directly [64]. Plus, THz pulse measurement was also used to estimate μ. David et al. used a two-color laser plasma in dry air to generate multi-THz pulses and excited the large $MAPbI_3$ single crystals and detected the electric field by an air-biased coherent detection scheme with 1–30 THz ultrabandwidth after normal incidence reflection off the crystal facet (**Figure 5e**, **f**) [65]. Such spectra measurements indicate the ultrafast dynamics and efficiencies of free charge creation and remarkably high μ as high as 500–800 cm^2/Vs. Furthermore, FETs are the fundamental components to realize digital integrated circuits, which are also often used as a platform to evaluate charge transport mechanism in the active materials. In this regard, bottom-gate, top-contact FETs were fabricated via using micrometer-thin $MAPbX_3$ (X = Cl, Br, and I) crystals as active layer (**Figure 4g**) [66], from which the field-effect μ values are up to 4.7 and 1.5 cm^2/Vs in p- and n-channel devices, respectively (**Figure 5h**).

Carrier lifetime (τ) is an important parameter that should be considered when designing an optoelectronic device. Upon excitation by photons, the active materials will be in an excited state. After that, the photo-induced holes and electrons will recombine back to the ground state. Usually, if this recombination process, that is, the carrier lifetime of carriers, is sufficiently long, a high performance device will be expected. The τ of semiconductors strongly depends on the nature, dimension, and purity of the materials. Generally, τ can be obtained from the PL decay, transient absorption, as well as the transient photo-voltage decay and impedance methods [69]. Among these methods, the PL decay approach has been widely applied. The superposition of fast and slow components of carrier dynamics from the PL spectra measurement result yield $\tau \approx 41$ and 357 ns for $MAPbBr_3$ (**Figure 5i**) [15, 70, 71]. Transient absorption (TA) also suggests the recombination property of excitons which is used to determine the carrier lifetime through a bi-exponential fitting [60]. The carrier diffusion length L_D can be further

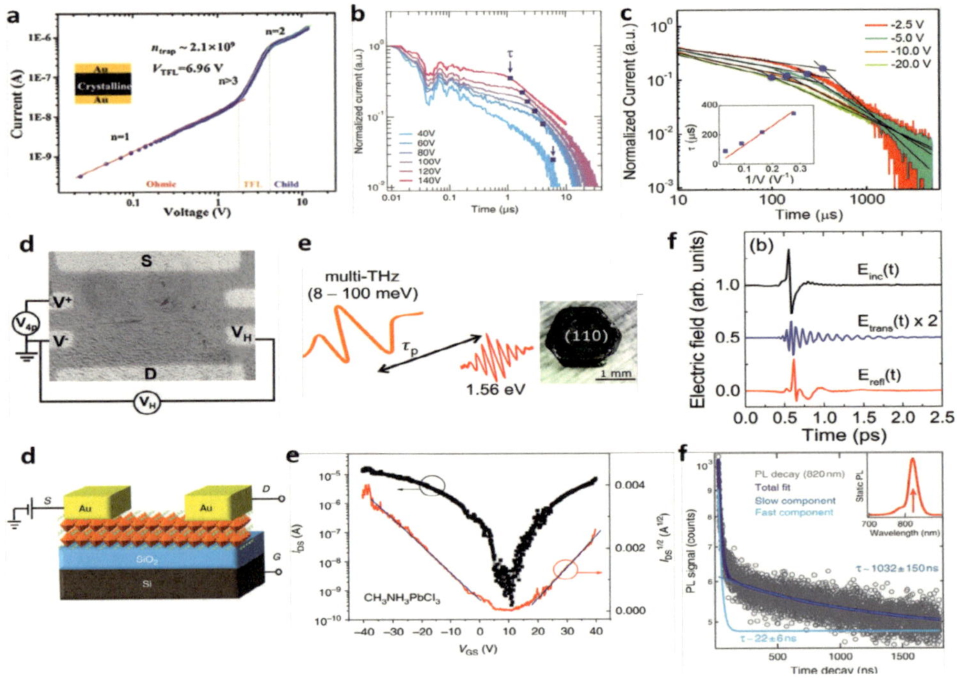

Figure 5.
5a, dark I-V *curve of hole-only $MAPbI_3$ crystal device.* J. Energy Chem. *[62], Copyright 2018.* ***5b**, ToF traces of $MAPbBr_3$ crystal.* Science *[15], copyright 2015.* ***5c**, transient current curves of perovskite crystal devices.* Science *[63], copyright 2015.* ***5d**, schematic of hall effect measurement.* Adv. Mater. *[64], Copyright 2016.* ***5e**, schematic of time-resolved multi-THz spectroscopy experiment.* **5f**, *incident (black), transmitted (blue) and reflected (red) multi-THz pulses after interaction with the crystal.* Energy Environ. Sci. *[65], Copyright 2015.* ***5 g**, schematic of bottom-gate, top-contact perovskite crystal FET.* ***5 h**, transfer characteristics of a $MAPbCl_3$ device.* Nature Commun. *[66], Copyright 2018.* ***5i**, PL time decay trace of a $MAPbBr_3$ crystal.* Science *[15], copyright 2015.*

estimated based on the equation: $L_D = [((k_B T)/e\mu\tau)]^{1/2}$, where k_B is Boltzmann's constant and T is the sample temperature. From the above-examined values of μ and τ, L_D was calculated [63, 64].

4. Applications of perovskite single crystals

4.1 Photovoltaic cells

The widely studied hybrid perovskite solar cells with high performance are usually made from polycrystalline films; however, the current studies have also focused on the developments and optimization of single crystal perovskite solar cells, owing to their significant advantages. Huang et al. fabricated photovoltaic devices based on $MAPbI_3$ bulk crystals by depositing gold (Au) as anodes and gallium (Ga) as cathodes (**Figure 6a**) [63]. A red-shift of 50 nm of the EQE cutoff to 850 nm showed that $MAPbI_3$ crystals increased the upper limit of short-circuit current density (*JSC*) compared with the polycrystalline solar cells from 27.5 mA/cm^2 to 33.0 mA/cm^2. Notably, as compared with the perovskite polycrystalline solar cells, the bulk crystal devices showed much lower efficiency, which was attributed to the fact that photogenerated carriers could not be fully collected in a thick active layer. Much thinner $MAPbBr_3$ monocrystalline films grown on indium tin oxide (ITO)-coated glass were applied into the solar cells, and the devices showed the best cell performance with a fill factor (*FF*) of 0.58, a *JSC* of 7.42 mA/cm^2, an open-circuit voltage (*VOC*) of

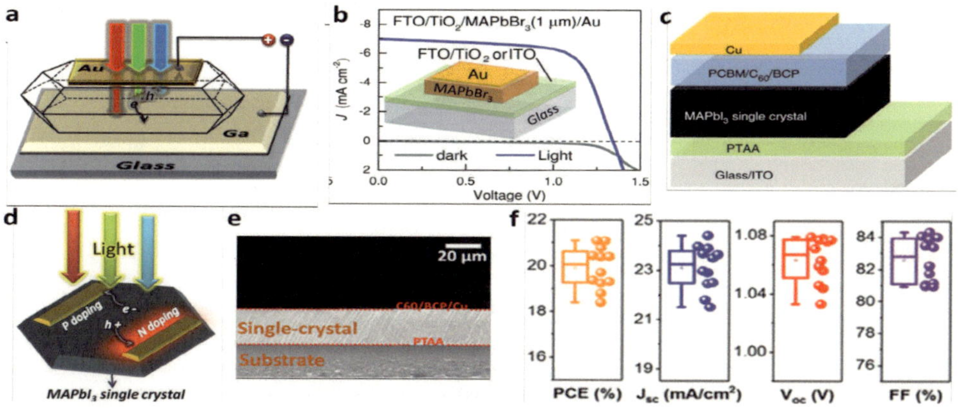

Figure 6.
***6a**, schematic of $MAPbI_3$ crystal solar cell.* Science *[63], copyright 2015.* ***6b**, dark and illuminated* J-V *curves of $MAPbBr_3$ crystal solar cells with a device illustration in the inset. Adv. Mater. [54], Copyright 2016.* ***6c**, device structure of single-crystal solar cells.* Nature Commun. *[72], Copyright 2017.* ***6d**, schematic of $MAPbI_3$ crystal solar cells with lateral structure.* Adv. Mater. *[73], Copyright 2016.* ***6e**, cross-sectional SEM image of a $MAPbI_3$ crystal device.* ***6f**, statistical summary of photovoltaic parameters from 12 devices.* ACS Energy Lett. *[74], Copyright 2019.*

1.24 V, and a *PCE* of 5.37% (**Figure 6b**) [54]. To enhance the device performance, Huang's group further fabricated crystal solar cells through interface engineering (**Figure 6c**), of which the best device showed a *JSC* of 20.5 mA/cm^2, a *VOC* of 1.06 V, a *FF* of 74.1%, and a *PCE* of 16.1% [72]. The single crystal solar cell also displayed the better device stability of remaining nearly unchanged after storage in air for 30 days.

In addition to the vertical-structured solar cells, Huang's group also fabricated the lateral structure perovskite crystal device (**Figure 6d**) [73], which showed a *VOC* of 0.82 V and the highest *PCE* of 5.36% at 170 K. More recently, a 20-μm $MAPbI_3$ single crystal inverted *p-i-n* solar cell with a *PCE* as high as 21.09% and a *FF* up to 84.3% was fabricated [74], of which the cross-sectional SEM image and photovoltaic performance are shown in **Figure 6e** and **f**. To further realize the optimized performance of perovskite crystal solar cells, more efforts will be performed to enhance the sample quality and to design promising device structures.

4.2 Photodetectors

Photodetectors which can convert incident light into electrical signals are critical for various industrial and scientific applications. To evaluate the photodetector performance, several parameters are important, including responsivity (*R*), detectivity (*D**), Gain (*G*), and linear dynamic range (*LDR*), which are listed and are explained in **Table 1** briefly.

4.2.1 In visible region

Huang's group fabricated perovskite crystal photodetectors that exhibited a high sensitivity capacity, which led to a narrow-band photo-response with a full width at half maximum (FWHM) of less than 20 nm under V = −1 V (**Figure 7a**) [49]. *EQE* spectra of the single crystals showed a narrow peak near the absorption edge, which promised a detection application at a specific wavelength, with a peak *D** over 2×10^{10} Jones at 570 nm under V = −4 V (**Figure 7b**). Also, Huang et al. further fabricated vertical structured perovskite crystal photodetectors by using the non-wetting hole transport layer-coating substrates [75]. The noise currents are as low as 1.4 and 1.8 fA/Hz$^{1/2}$ at an 8-Hz frequency for the devices based on $MAPbBr_3$

Quantity	Unit	Definition
Photocurrent (I_{light})	A	Current through a photodetector resulting from illumination.
Dark-current (I_{dark})	A	Current through a device in the absence of illumination.
Photoresponsivity (R)	A/W	R is calculated according to: $R = (I_{light} - I_{dark})/P_{light}$, where P_{light} is power of the incident light.
Detectivity (D^*)	Jones	D^* can be calculated as $R/(2eJ_d)^{1/2}$, where e is elementary charge and J_d is dark current density.
Gain (G)	—	G can be calculated as $[(J_{light} - J_{dark})/e]/(P_{light}/h\nu)$, where $h\nu$ is the incident photon energy.
Linear dynamic range (LDR)	dB	LDR is calculated by $LDR = 20\log(P_{sat}/P_{low})$, where P_{sat} (P_{low}) is the light intensity when the incident light intensity stronger (weaker) than which the photocurrent begins to deviate from linearity.
External quantum efficiency (EQE)	%	Carrier number divided by the number of incident photons.
Internal quantum efficiency (IQE)	%	It is the ratio of carrier number to the number of incident photons that are absorbed by the device.

Table 1.
Parameters for evaluating the perovskite single crystal photodetectors.

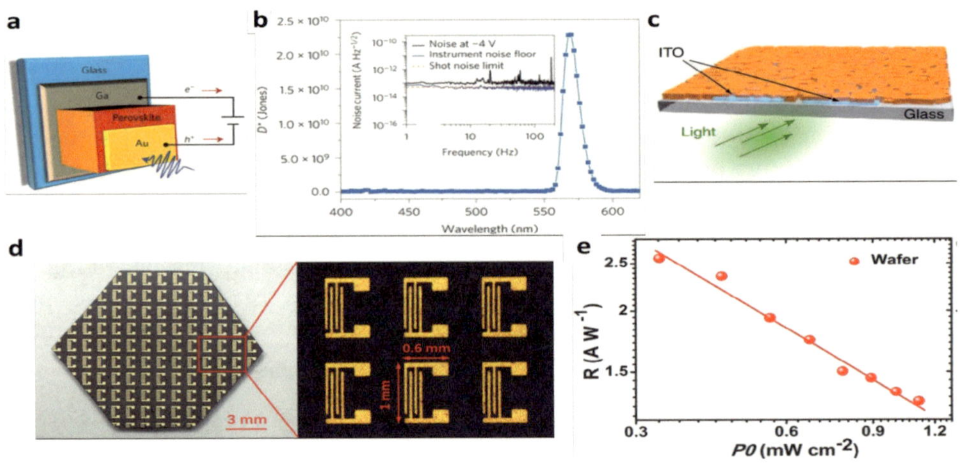

Figure 7.
7a, schematic of device structure. 7b, D* *spectrum and total noise at −4 V.* Nature Photonics *[49], copyright 2015. 7c, illustration of planar-integrated $MAPbBr_3$ photodetector.* Nature Commun. *[42], Copyright 2015. Photograph of ≈100 photodetectors on a perovskite crystal wafer (7d) and the* R *values (7e).* Adv. Mater. *[55], Copyright 2016.*

and $MAPbI_3$, respectively. Additionally, the photocurrent responses of both the $MAPbBr_3$ and $MAPbI_3$ devices were linear, and their *LDR*s are up to 256 and 222 dB, respectively. Sun's group introduced a planar-type photodetector on the $MAPbI_3$ crystal (001) facet with a highest *R* value of 953 A/W and *EQE* of 2.22×10^5% at a light power density of 2.12 nW/cm^2 [76]. Wei's group used a two-step method to fabricate a self-powered photodetector based on a $MAPbBr_3$ crystal core-shell heterojunction [77]. The device showed a broad photo-response ranging from 350 to 800 nm and a peak *R* up to 11.5 mA/W. Hu's group fabricated photodetectors based on $MAPbI_3$ single crystal nanowires and nanoplates by transferring them to SiO_2/Si slides [78]. The highest On/Off ratio approached 10^3 under a light illumination of 73.7 mW/cm^2.

Although perovskite crystal photodetectors have shown better performance, macroscopic crystals cannot be grown on a planar substrate, restricting their potential for device integration. To overcome this shortcoming, Bakr et al. grew large-area planar-integrated crystal films onto the ITO-patterned substrates (**Figure 7c**) [42], and the fabricated photodetector possessed a high G (above 10^4) and a high gain-bandwidth product (above 10^8 Hz) relative to other perovskite devices. Furthermore, Liu's group fabricated a photodetector based on a thin perovskite crystal wafer by the space-limited crystallization method, which has about 100 pairs of interdigitated Au wire electrodes (**Figure 7d**) [55], and the R increased linearly as the radiance intensity decreased (**Figure 7e**). Moreover, Su's group sputtered the thin Au electrodes on a large-area $MAPbBr_3$ thin crystal to fabricate a narrowband photodetector [56]. Furthermore, Ma's group reported the superior-performance photodetectors based on $MAPbBr_3$ thin crystals [79], which displayed the R as high as 1.6×10^7 A/W and the highest G up to 5×10^7.

4.2.2 In ultraviolet (UV) region

UV detection is a key technology in the fields of flame detection [80], remote security monitoring [81], environmental monitoring [82], and so forth. Researchers have endeavored to develop UV photodetectors based on perovskite crystals considering their excellent UV absorption properties. Visible-blind UV photodetectors based on $MAPbCl_3$ crystals a suitable bandgap of about 3.11 eV were fabricated (**Figure 8a**) [60], and the device showed the dark current as low as 4.15×10^{-7} A at 15 V and a drastically high stability (**Figure 8b**). Planar-integrated $MAPbCl_3$ crystal UV photodetectors on ITO-deposited glass substrate were reported by Sargent et al. (**Figure 8c**) [83], which showed decreased R and G values as increased power density of a 385-nm laser (**Figure 8d**) [85].

4.2.3 In near-infrared (NIR) region

NIR photodetectors have widespread uses in telecommunications [86], as well as thermal and biological imaging [87–90]. Meredith's group demonstrated the perovskite crystal that overcame the large bandgap and presented photodetectors with performance metrics appropriate for NIR detection by using the trap-related linear sub-gap absorption (**Figure 8e**) [84]. A strong NIR photo-response was achieved in photodiodes based on $MAPbI_3$ crystals illuminated by a continuous 808-nm laser (~10 mW/cm^2). The photodiodes could also respond to a laser with a wavelength as long as 1064 nm (**Figure 8f**).

4.2.4 In X-ray region

In addition to the common light detections from UV to IR, perovskite crystals have been employed for the detection of X-rays, which have important applications in medical diagnostics, clinical treatment, and the non-destructive testing of products [53]. Huang et al. fabricated a sensitive $MAPbBr_3$ crystal X-ray detector with the structure of Au/$MAPbBr_3$/crystal/C_{60}/BCP/Ag or Au (**Figure 9a**) [53]. Through reducing the bulk defects and passivating surface traps, the devices showed a detection efficiency of 16.4% at a near zero bias under irradiation with continuum X-ray energy up to 50 keV. The lowest detectable X-ray dose rate was 0.5 $\mu Gy_{air}/s$ with a sensitivity of 80 $\mu C/Gy_{air}cm^2$, which is four times higher than the sensitivity achieved in α-Se-based X-ray detectors (**Figure 9b**). An X-ray detector based on *p-i-n* diode array made of a thick $MAPbBr_3$ single crystal was introduced by Chen's group [94], which displayed the highest sensitivity of 23.6 $\mu C/mGy_{air}cm^2$, indicating high potential for practical applications.

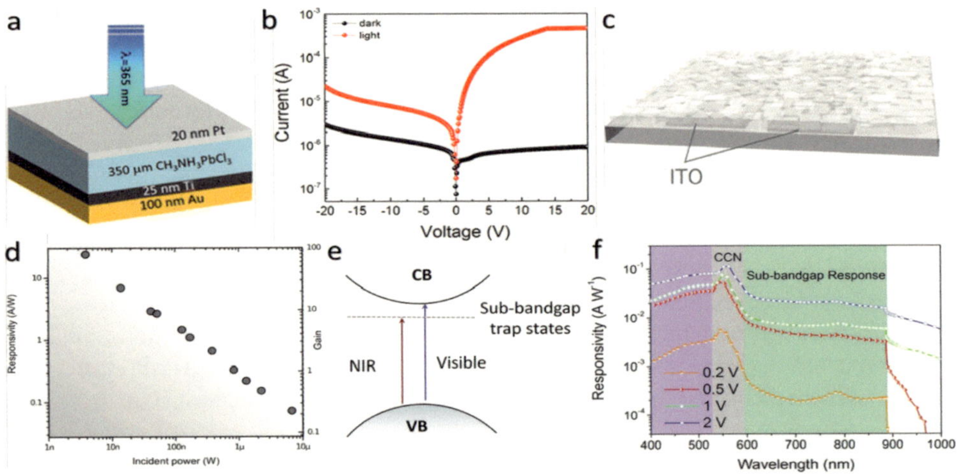

Figure 8.
***8a**, device architecture of $MAPbCl_3$ crystal photodetector. **8b**,* I-V *curves of the photodetector under UV light (λ = 365 nm) and in the dark.* J. Phys. Chem. Lett. *[60], Copyright 2015. **8c**, schematic of planar-integrated $MAPbCl_3$ UV-detectors. **8d**,* R *and* G *values vs. incident light power.* Adv. Mater. *[83], Copyright 2016. **8e**, sub-gap electron trap state absorptions. **8f**,* R *values of $MAPbI_3$ photo-resistors under the illumination above the gap (visible, 600 nm) and below the gap (NIR, 900 nm).* Laser Photonics Rev. *[84], copyright 2016.*

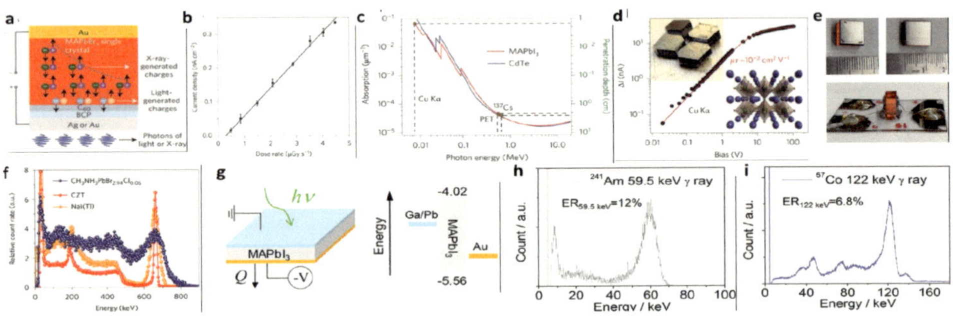

Figure 9.
***9a**, structure of $MAPbBr_3$ crystal X-ray detector. **9b**, X-ray-generated photocurrent at various dose rates.* Nature Photonics *[53], copyright 2016. **9c**, attenuation coefficient and penetration depth of $MAPbI_3$ and CdTe. **9d**, photocurrent and a fit with Hecht model generated by Cu Kα X-ray radiation (8 keV) in a $MAPbI_3$ crystal.* Nature Photonics *[91], copyright 2016. **9e**, pictures of guard ring electrode side, anode side and side view of a $MAPbBr_{2.94}Cl_{0.06}$ crystal detector. **9f**, ^{137}Cs energy spectrum obtained by crystal, CZT and NaI (Tl) detectors.* Nature Mater. *[92], Copyright 2017. **9 g**, schematic of a Schottky-type $MAPbI_3$ detector with asymmetrical electrode and the energy level diagram. Energy-resolved spectrum by Schottky-type $MAPbI_3$ detector (**9 h**) under ^{241}Am 59.5 keV γ-ray under −50 V and (**9i**) under ^{57}Co 122 keV γ-ray under −70 V.* ACS Photonics *[93], copyright 2018.*

4.2.5 In gamma-ray (γ-ray) region

Similar to X-ray detectors, the γ-ray detectors are also widely used in many fields, owing to the non-invasive detections. However, γ-ray detectors need to work in a weak radiation field pulse mode and perform event-by-event detections to sort out the intensity vs. the energy of the radiation quanta. Large and balanced μ and τ are needed for high-energy detection. Huang et al. reported high-quality $MAPbI_3$ crystals that were applied to γ-ray detection with a 4% efficiency when operating in the γ-voltaic mode [63]. Kovalenko et al. demonstrated $MAPbI_3$ crystals for γ-ray detection (**Figure 9c**), and a 59.6 keV ^{241}Am energy spectrum was acquired [91]. A fit of bias dependence of photocurrent with Hecht model indicated a high $\mu\tau$ product of ~10^{-2} cm^2/V (**Figure 9d**) [95, 96].

Huang's group further reported a Cl^- dopant compensation of $MAPbBr_3$ single crystal process to fabricate a low-cost γ-ray detector [92]. $MAPbBr_{2.94}Cl_{0.06}$ crystals with a larger $\mu\tau$ product were equipped with a guard ring electrode to mitigate their leakage current (**Figure 9e**). The ^{137}Cs energy spectrum obtained by such crystals with a full-energy peak resolution of 6.50% is compared with the spectrum obtained by CZT and NaI(Tl) detectors (**Figure 9f**). A high-performance $MAPbI_3$ crystal γ-ray spectrometer was designed by Kanatzidis et al. [93], and the asymmetrical electrodes (Schottky-type) were applied to prohibit the hole injection from the anode or to reduce the leakage current (**Figure 9g**). The best energy resolution of the device for ^{241}Am 59.5 keV γ-rays was ~12%; while the best energy resolution achieved for ^{57}Co 122 keV was 6.8% (**Figure 9h** and **i**).

4.3 Light-emitting diodes (LEDs) and lasers

With the exceptional PL efficiency and high color purity, perovskite crystals can also perform as high-performance LEDs [97]. Most of the existing perovskite LEDs employ a polycrystalline film with sizes of nanometers to micrometers, and coherent light emission is a challenge [98]. In Yu's work, the LEDs with the structure of ITO/$MAPbBr_3$ micro-platelet/Au cathode had the turn-on voltage of about 1.8 V and could last for at least 54 h with a luminance of ~5000 cd/m^2 (**Figure 10a**) [99].

The excellent properties, including a small trap density, long lifetime and electron–hole diffusion length, and large carrier mobility, also make perovskite crystals suitable for laser devices with low lasing thresholds and high qualities. Xiong's group grew typical $MAPbI_3$ triangular nano-platelets and optically pumped them by a femtosecond-pulsed laser (**Figure 10b**) [100], and the peaks centered at λ = 776.7, 779.2, 781.9, 784.3, and 786.8 nm appeared over the spontaneous emission band with a FWHM of ~1.2 nm (**Figure 10c**), when the pump fluence was increased to 40.6 μJ/cm^2. Zhu et al. demonstrated room-temperature lasing via using $MAPbI_3$ crystal nanowire, which had a broad tenability covering the NIR to visible region [101].

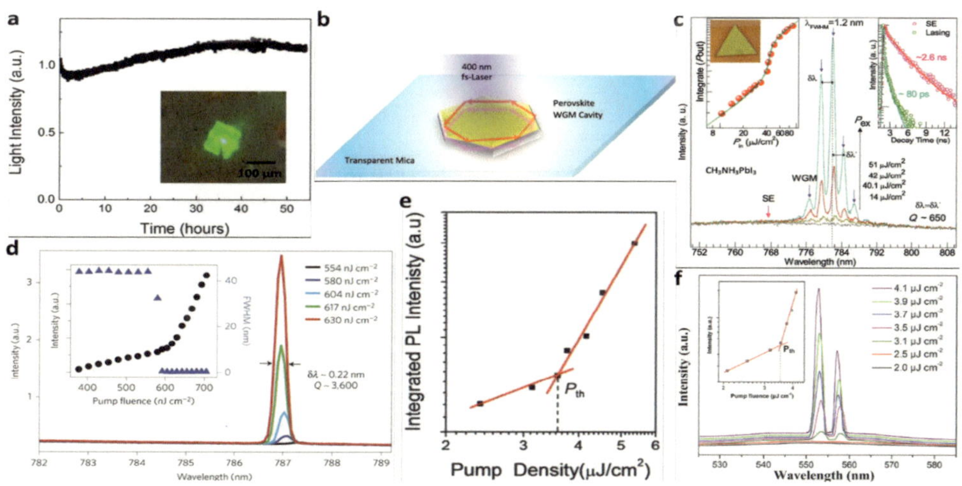

Figure 10.
10a, light emission intensity vs. time of a perovskite LED at −193°C. inset: A microscopic image at t = 12 h. ACS Nano *[99], copyright 2018.* ***10b****, schematic for optical setup of a $CH_3NH_3PbI_3$ nanoplatelet.* ***10c****, evolution from spontaneous emission to lasing in a typical $CH_3NH_3PbI_3$ nanoplatelet. Inset left: Optical image of a nanoplate and plot of integrated P_{out}. Inset right: PL decay curve below (pink) and above (dark green) the threshold.* Nano Lett. *[100], Copyright 2014.* ***10d****, nanowire emission spectra. Inset: Integrated emission intensity and FWHM vs.* P. Nature Mater. *[101], Copyright 2015.* ***10e****, integrated PL intensity as a function of excitation density.* Adv. Mater. *[102], Copyright 2015.* ***10f****, emission spectra of perovskite microplates excited by different pump densities. Inset: Integrated PL intensity vs. pump density.* Adv. Mater. *[103], Copyright 2016.*

From **Figure 10d**, a sharp peak appeared at 787 nm in the representative emission spectra and grew rapidly with increasing the pump laser fluence (P) with the lasing threshold P_{Th} of ~595 nJ/cm^2. Additionally, $MAPbBr_3$ crystal square micro-disks were synthesized into a 557-nm single-mode laser based on a built-in whispering gallery mode micro-resonator by Fu's group [102], from which a P_{Th} = 3.6 μJ/cm^2 was observed, and a sublinear regime was observed below the threshold (**Figure 10e**). Uniform-sized $MAPbBr_3$ microplates were also created by Jiang et al. by using "liquid knife" and were made into lasers [103]. A 400-nm pulsed laser beam was used as a pump source to excite microplates, and a spontaneous emission peak centered at 550 nm with a FWHM of 20 nm was observed (**Figure 10f**).

5. Challenges and open issues

Hybrid perovskite single crystals have shown great potential in high-performance optoelectronic devices; however, several challenges and issues still remain in terms of their practical applications. They mainly include (1) the effects of surface defects, (2) the large-area fabrication, as well as (3) the stability of the perovskite single crystal devices and (4) the health and environmental concerns.

5.1 Surface defects

The absence of grain boundaries makes perovskite crystals acquire better optical and charge transport properties than their polycrystalline counterparts. However, the surface of crystals usually possesses lots of chemical impurities, dangling bonds, surface dislocations, and under-coordinated atoms, and becomes disordered owing to hydration, thus decreasing the carrier mobility and carrier diffusion length and promote the recombination of carriers near the crystal surface [76, 104–106]. Thus, the further decrease of defects, especially the surface defects, is required, aiming to gain high-quality perovskite crystals. To realize high-performance optoelectronic devices based on perovskite crystals with low-level surface defects, more research should be carried out on the surface passivation.

5.2 Large-area fabrication

Hybrid perovskite thin crystals are freer of grain boundaries and exhibit better transport properties than those of the polycrystalline candidates, so their large-area fabrication will ensure a promising future. However, the embedding of volatile and vulnerable organic components on fragile inorganic framework makes them difficult to be fabricated with a large area by deposition techniques or solution-based methods [42, 54]. Furthermore, thin crystals were grown directly on conductive substrates like FTO- or ITO-glass [42, 56], and tailored substrates, such as SiO_2/Si [97], which provide in-situ growth for thin crystals and be directly made into devices. Nevertheless, these large-area thin crystals have rough surfaces and a great number of surface defects, and thus their optoelectronic properties remain inferior to the bulk counterparts. Further optimization of growth methods for large-area thin crystals is needed for industry productions in future.

5.3 Long-term stability

Low stability of the current hybrid perovskite crystal devices hinders their broad practical application. Several factors that affect the device stability, like ion migration [107, 108], can cause hysteresis and photo-induced phase separation, and the interaction between single crystals and their surroundings lies in the degradation of perovskite by humidity and light [109–111]. Therefore, to further enhance the stability of single crystal devices, optimized device structures should be designed to control the ion migrations. Meanwhile, various compositions and interface engineering approaches are also intensively investigated to confront this critical issue. In addition, encapsulation has been demonstrated to be a valid method to protect hybrid perovskite devices.

5.4 Health and environmental concerns

The growth of hybrid perovskite crystals adopt heavy metal ions, like lead (Pb) or tin (Sn), and organic functional groups, which can impact both the environment and human health. This critical issue needs to be overcome, aiming for further commercialization. As for the common $MAPbI_3$ perovskite crystal, the Pb-ion is toxic to both the human health and natural environment; while the organic solvents used during the growth process of crystals are also toxic and easily penetrate into the human body [112]. To solve these problems, capsulation and recycling are needed in the use of crystal materials and organic solvents. Furthermore, other alternative metals to Pb, with a lower toxicity, are also being studied, such as bismuth and antimony [113, 114], and thus, the optoelectronic properties of these Pb-free perovskite crystals need to be explored further for device applications.

6. Conclusions

More recently, hybrid perovskite crystals, having different dimensional forms: bulk and thin crystals, and micro−/nano-plates, have been widely explored as functional layers for optoelectronic devices owing to their excellent physical properties combined with the advantage of ease of processing. Although these types of devices are still in the early stages of development, a strong potential for a variety of technological and commercial applications clearly remains. Here, we presented a comprehensive overview of the recent advances in hybrid perovskite crystals with respect to the background knowledge on the optoelectronic properties and charge transport dynamics of crystals, and their applications in the area of optoelectronic devices, and a fundamental understanding of the device performance. We summarized the main growth methods for the bulk crystals and also some modified and optimized approaches to synthesize thin crystals. The detailed discussions are focused on charge transport characteristics, operation mechanisms, and challenges, aiming to provide a critical understanding of further advance in materials design and device engineering in a variety of optoelectronic technologies.

In conclusion, the research progress achieved to date in the area of perovskite crystal optoelectronic devices, with the emphasis placed on challenges faced by the research community, has been summarized systematically, and finally perspective on the opportunities offered by this emerging family to photoactive materials in practical and commercial technologies is also proposed. Further exploration of high-quality perovskite crystals, combined with an in-depth understandings of working mechanism of devices, indicates a promising future for wide applications with markedly-enhanced performance.

Acknowledgements

The author acknowledges support from Discovery Early Career Researcher Award (DECRA) (DE180100167) from the Australian Research Council (ARC).

Conflict of interest

There are no conflicts to declare.

Author details

Feng Li
School of Physics, Faculty of Science, The University of Sydney, Sydney, NSW, Australia

*Address all correspondence to: feng.li2@sydney.edu.au

References

[1] Mitzi, D. B. Templating and structural engineering in organic–inorganic perovskites. *J. Chem. Soc., Dalton Trans.*, 2001, 1-12.

[2] Green, M. A.; Ho-Baillie, A.; Snaith, H. J. The emergence of perovskite solar cells. *Nature photonics* 2014, 8, 506.

[3] Zhang, D.; Zhu, Y.; et al. Atomic-resolution transmission electron microscopy of electron beam–sensitive crystalline materials. *Science* 2018, 359, 675-679.

[4] Ke, W.; Mao, L.; et al. Compositional and solvent engineering in Dion–Jacobson 2D perovskites boosts solar cell efficiency and stability. *Adv. Energy Mater.* 2019, 9, 1803384.

[5] Li, S.; Tong, S.; et al. Fast-response and high-responsivity $FA_xMA_{(1-x)}PbI_3$ photodetectors fabricated via doctor-blading deposition in ambient condition. *Org. Electron.* 2018, 52, 190-194.

[6] Wang, H.; Wu, H.; et al. Controllable $Cs_xFA_{1-x}PbI_3$ single-crystal morphology via rationally regulating the diffusion and collision of micelles toward high-performance photon detectors. *ACS Appl. Mater. Interfaces* 2019, 11, 13812-13821.

[7] Hou, X.; Xu, M.; et al. High performance printable perovskite solar cells based on $Cs_{0.1}FA_{0.9}PbI_3$ in mesoporous scaffolds. *J. Power Sources* 2019, 415, 105-111.

[8] Wang, M.; Jiang, X.; et al. High-performance and stable mesoporous perovskite solar cells via well-crystallized $FA_{0.85}MA_{0.15}Pb(I_{0.8}Br_{0.2})_3$. *ACS Appl. Mater. Interfaces* 2018, 11, 2989-2996.

[9] Babayigit, A.; Thanh, D. D.; et al. Assessing the toxicity of Pb- and Sn-based perovskite solar cells in model organism *Danio rerio. Sci. Rep.* 2016, 6, 1-11.

[10] Ma, H.; Ma, Y.; et al. Experimental phonon dispersion and lifetimes of tetragonal $CH_3NH_3PbI_3$ perovskite crystals. *J. Phys. Chem. Let.* 2018, 10, 1-6.

[11] Ju, D.; Zheng, X.; et al. Reversible band gap narrowing of Sn-based hybrid perovskite single crystal with excellent phase stability. *Angew. Chem. Int. Ed.* 2018, 57, 14868-14872.

[12] Ju, D.; Dang, Y.; et al. Tunable band gap and long carrier recombination lifetime of stable mixed $CH_3NH_3Pb_xSn_{1-x}Br_3$ single crystals. *Chem. Mater.* 2018, 30, 1556-1565.

[13] Shao, S.; Liu, J.; et al. Highly reproducible Sn-based hybrid perovskite solar cells with 9% efficiency. *Adv. Energy Mater.* 2018, 8, 1702019.

[14] Li, C.; Lu, X.; et al. Formability of ABX_3 (X= F, Cl, Br, I) halide perovskites. *Acta Crystallogr. B Struct. Sci.* 2008, 64, 702-707.

[15] Shi, D.; Adinolfi, V.; et al. Low trap-state density and long carrier diffusion in organolead trihalide perovskite single crystals. *Science* 2015, 347, 519-522.

[16] Kojima, A.; Teshima, K.; et al. Organometal halide perovskites as visible-light sensitizers for photovoltaic cells. *J. Am. Chem. Soc.* 2009, 131, 6050-6051.

[17] Xiao, Z.; Bi, C.; et al. Efficient, high yield perovskite photovoltaic devices grown by interdiffusion of solution-processed precursor stacking layers. *Energy Environ. Sci.* 2014, 7, 2619-2623.

[18] Liu, M.; Johnston, M. B.; Snaith, H. J. Efficient planar heterojunction perovskite solar cells by vapour deposition. *Nature* 2013, 501, 395-398.

[19] Xiao, M.; Huang, F.; et al. A fast deposition-crystallization procedure for highly efficient lead iodide perovskite thin-film solar cells. *Angew. Chem. Int. Ed.* 2014, 53, 9898-9903.

[20] Zhu, W.; Bao, C.; et al. An efficient planar-heterojunction solar cell based on wide-bandgap $CH_3NH_3PbI_{2.1}Br_{0.9}$ perovskite film for tandem cell application. *Chem. Commun.* 2016, 52, 304-307.

[21] Heo, J. H.; Im, S. H.; et al. Efficient inorganic–organic hybrid heterojunction solar cells containing perovskite compound and polymeric hole conductors. *Nature photonics* 2013, 7, 486.

[22] Jeon, N. J.; Noh, J. H.; et al. Solvent engineering for high-performance inorganic–organic hybrid perovskite solar cells. *Nature Mater.* 2014, 13, 897-903.

[23] Lee, M. M.; Teuscher, J.; et al. Efficient hybrid solar cells based on meso-super structured organometal halide perovskites. *Science* 2012, 338, 643-647.

[24] Filip, M. R.; Eperon, G. E.; et al. Steric engineering of metal-halide perovskites with tunable optical band gaps. *Nature Commun.* 2014, 5, 1-9.

[25] D'Innocenzo, V.; Srimath Kandada, A. R.; et al. Tuning the light emission properties by band gap engineering in hybrid lead halide perovskite. *J. Am. Chem. Soc.* 2014, 136, 17730-17733.

[26] Sum, T. C.; Mathews, N. Advancements in perovskite solar cells: photophysics behind the photovoltaics. *Energy Environ. Sci.* 2014, 7, 2518-2534.

[27] Choi, H.; Jeong, J.; et al. Cesium-doped methylammonium lead iodide perovskite light absorber for hybrid solar cells. *Nano Energy* 2014, 7, 80-85.

[28] Kim, Y. H.; Cho, H.; et al. Multicolored organic/inorganic hybrid perovskite light-emitting diodes. *Adv. Mater.* 2015, 27, 1248-1254.

[29] NREL. Efficiency chart. https://www.nrel.gov/pv/assets/pdfs/pv-efficiency-chart.20181214.pdf.

[30] Hu, X.; Zhang, X.; et al. High-performance flexible broadband photodetector based on organolead halide perovskite. *Adv. Funct. Mater.* 2014, 24, 7373-7380.

[31] Chin, X. Y.; Cortecchia, D.; et al. Lead iodide perovskite light-emitting field-effect transistor. *Nature Commun.* 2015, 6, 7383.

[32] Li, F.; Ma, C.; et al. Ambipolar solution-processed hybrid perovskite phototransistors. *Nature Commun.* 2015, 6, 8238.

[33] Meng, L.; Yao, E. P.; et al. Pure formamidinium-based perovskite light-emitting diodes with high efficiency and low driving voltage. *Adv. Mater.* 2017, 29, 1603826.

[34] Liu, P.; He, X.; et al. Organic–inorganic hybrid perovskite nanowire laser arrays. *ACS Nano* 2017, 11, 5766-5773.

[35] Duan, H.-S.; Zhou, H.; et al. The identification and characterization of defect states in hybrid organic–inorganic perovskite photovoltaics. *Phys. Chem. Chem. Phys.* 2015, 17, 112-116.

[36] Zheng, X.; Chen, B.; et al. Defect passivation in hybrid perovskite solar cells using quaternary ammonium halide anions and cations. *Nature Energy* 2017, 2, 1-9.

[37] Liu, Z.; Hu, J.; et al. Chemical reduction of intrinsic defects in thicker heterojunction planar perovskite solar cells. *Adv. Mater.* 2017, 29, 1606774.

[38] Fang, H.-H.; Adjokatse, S.; et al. Ultrahigh sensitivity of methylammonium lead tribromide perovskite single crystals to environmental gases. *Sci. Adv.* 2016, 2, e1600534.

[39] Huang, J.; Shao, Y.; Dong, Q. Organometal trihalide perovskite single crystals: a next wave of materials for 25% efficiency photovoltaics and applications beyond? *J. Phys. Chem. Lett.* 2015, 6, 3218-3227.

[40] Zhang, F.; Yang, B.; et al. Extra-long electron–hole diffusion lengths in $CH_3NH_3PbI_{3-x}Cl_x$ perovskite single crystals. *J. Mater. Chem. C* 2017, 5, 8431-8435.

[41] Chen, B.-X.; Li, W.-G.; et al. Large-grained perovskite films via $FA_xMA_{1-x}Pb(I_xBr_{1-x})_3$ single crystal precursor for efficient solar cells. *Nano Energy* 2017, 34, 264-270.

[42] Saidaminov, M. I.; Adinolfi, V.; et al. Planar-integrated single-crystalline perovskite photodetectors. *Nature Commun.* 2015, 6, 1-7.

[43] Li, S.; Zhang, C.; et al. Metal halide perovskite single crystals: from growth process to application. *Crystals* 2018, 8, 220.

[44] Liu, Y.; Yang, Z.; Liu, S. Recent progress in single-crystalline perovskite research Including crystal preparation, property evaluation, and applications. *Adv. Sci.* 2018, 5, 1700471.

[45] Chen, Y.; He, M.; et al. Structure and growth control of organic–inorganic halide perovskites for optoelectronics: from polycrystalline films to single crystals. *Adv. Sci.* 2016, 3, 1500392.

[46] Ding, J.; Yan, Q. Progress in organic-inorganic hybrid halide perovskite single crystal: growth techniques and applications. *Sci. China Mater.* 2017, 60, 1063-1078.

[47] Dang, Y.; Liu, Y.; et al. Bulk crystal growth of hybrid perovskite material $CH_3NH_3PbI_3$. *CrystEngComm* 2015, 17, 665-670.

[48] Su, J.; Chen, D.; Lin, C. Growth of large $CH_3NH_3PbX_3$ (X= I, Br) single crystals in solution. *J. Cryst. Growth* 2015, 422, 75-79.

[49] Fang, Y.; Dong, Q.; et al. Highly narrowband perovskite single-crystal photodetectors enabled by surface-charge recombination. *Nature Photonics* 2015, 9, 679.

[50] Saidaminov, M. I.; Abdelhady, A. L.; et al. High-quality bulk hybrid perovskite single crystals within minutes by inverse temperature crystallization. *Nature Commun.* 2015, 6, 1-6.

[51] Zhang, Y.; Liu, Y.; et al. Perovskite $CH_3NH_3Pb(Br_xI_{1-x})_3$ single crystals with controlled composition for fine-tuned bandgap towards optimized optoelectronic applications. *J. Mater. Chem. C* 2016, 4, 9172-9178.

[52] Zhang, B.; Yan, J.; et al. Effect of the modulating of organic content on optical properties of single-crystal perovskite. *Opt. Mater.* 2016, 62, 273-278.

[53] Wei, H.; Fang, Y.; et al. Sensitive X-ray detectors made of methylammonium lead tribromide perovskite single crystals. *Nature Photonics* 2016, 10, 333.

[54] Peng, W.; Wang, L.; et al. Solution-grown monocrystalline hybrid perovskite films for hole-transporter-free solar cells. *Adv. Mater.* 2016, 28, 3383-3390.

[55] Liu, Y.; Zhang, Y.; et al. Thinness-and shape-controlled growth for

ultrathin single-crystalline perovskite wafers for mass production of superior photoelectronic devices. *Adv. Mater.* 2016, 28, 9204-9209.

[56] Rao, H. S.; Li, W. G.; et al. In-situ growth of 120-cm^2 $CH_3NH_3PbBr_3$ perovskite crystal film on FTO glass for narrowband-photodetectors. *Adv. Mater.* 2017, 29, 1602639.

[57] Chen, Y.-X.; Ge, Q.-Q.; et al. General space-confined on-substrate fabrication of thickness-adjustable hybrid perovskite single-crystalline thin films. *J. Am. Chem. Soc.* 2016, 138, 16196-16199.

[58] Lee, L.; Baek, J.; et al. Wafer-scale single-crystal perovskite patterned thin films based on geometrically-confined lateral crystal growth. *Nature Commun.* 2017, 8, 1-8.

[59] Gu, Z.; Huang, Z.; et al. A general printing approach for scalable growth of perovskite single-crystal films. *Sci. Adv.* 2018, 4, eaat2390.

[60] Maculan, G.; Sheikh, A. D.; et al. $CH_3NH_3PbCl_3$ single crystals: inverse temperature crystallization and visible-blind UV-photodetector. *J. Phys. Chem. Lett.* 2015, 6, 3781-3786.

[61] Wenger, B.; Nayak, P. K.; et al. Consolidation of the optoelectronic properties of $CH_3NH_3PbBr_3$ perovskite single crystals. *Nature Commun.* 2017, 8, 1-10.

[62] Zhang, Y.; Liu, Y.; et al. High-quality perovskite $MAPbI_3$ single crystals for broad-spectrum and rapid response integrate photodetector. *J. Energy Chem.* 2018, 27, 722-727.

[63] Dong, Q.; Fang, Y.; et al. Electron-hole diffusion lengths >175 μm in solution-grown $CH_3NH_3PbI_3$ single crystals. *Science* 2015, 347, 967-970.

[64] Yi, H. T.; Wu, X.; et al. Intrinsic charge transport across phase transitions in hybrid organo-inorganic perovskites. *Adv. Mater.* 2016, 28, 6509-6514.

[65] Valverde-Chávez, D. A.; Ponseca, C. S.; et al. Intrinsic femtosecond charge generation dynamics in single crystal $CH_3NH_3PbI_3$. *Energy Environ. Sci.* 2015, 8, 3700-3707.

[66] Yu, W.; Li, F.; et al. Single crystal hybrid perovskite field-effect transistors. *Nature Commun.* 2018, 9, 1-10.

[67] Edri, E.; Kirmayer, S.; et al. High open-circuit voltage solar cells based on organic–inorganic lead bromide perovskite. *J. Phys. Chem. Lett.* 2013, 4, 897-902.

[68] Giorgi, G.; Yamashita, K. Organic–inorganic halide perovskites: an ambipolar class of materials with enhanced photovoltaic performances. *J. Mater. Chem. A* 2015, 3, 8981-8991.

[69] Ball, J. M.; Petrozza, A., Defects in perovskite-halides and their effects in solar cells. *Nature Energy* 2016, 1, 1-13.

[70] Zhang, M.; Yu, H.; et al. Composition-dependent photoluminescence intensity and prolonged recombination lifetime of perovskite $CH_3NH_3PbBr_{3-x}Cl_x$ films. *Chem. Commun.* 2014, 50, 11727-11730.

[71] Fang, Y.; Wei, H.; et al. Quantification of re-absorption and re-emission processes to determine photon recycling efficiency in perovskite single crystals. *Nature Commun.* 2017, 8, 14417.

[72] Chen, Z.; Dong, Q.; et al. Thin single crystal perovskite solar cells to harvest below-bandgap light absorption. *Nature Commun.* 2017, 8, 1-7.

[73] Dong, Q.; Song, J.; et al. Lateral-structure single-crystal hybrid perovskite solar cells via piezoelectric poling. *Adv. Mater.* 2016, 28, 2816-2821.

[74] Chen, Z.; Turedi, B.; et al. Single-crystal $MAPbI_3$ perovskite solar cells exceeding 21% power conversion efficiency. *ACS Energy Lett.* 2019, 4, 1258-1259.

[75] Bao, C.; Chen, Z.; et al. Low-noise and large-linear-dynamic-range photodetectors based on hybrid-perovskite thin-single-crystals. *Adv. Mater.* 2017, 29, 1703209.

[76] Lian, Z.; Yan, Q.; et al. High-performance planar-type photodetector on (100) facet of $MAPbI_3$ single crystal. *Sci. Rep.* 2015, 5, 16563.

[77] Cao, M.; Tian, J.; et al. Perovskite heterojunction based on $CH_3NH_3PbBr_3$ single crystal for high-sensitive self-powered photodetector. *Appl. Phys. Lett.* 2016, 109, 233303.

[78] Qin, X.; Yao, Y.; et al. Perovskite photodetectors based on $CH_3NH_3PbI_3$ single crystals. *Chem. Asian J.* 2016, 11, 2675-2679.

[79] Yang, Z.; Deng, Y.; et al. High-performance single-crystalline perovskite thin-film photodetector. *Adv. Mater.* 2018, 30, 1704333.

[80] Cheong, P.; Chang, K.-F.; et al. A ZigBee-based wireless sensor network node for ultraviolet detection of flame. *IEEE Trans. Ind. Electron.* 2011, 58, 5271-5277.

[81] Li, M.; Li, W.; et al. On-site determination and monitoring of real-time fluence delivery for an operating UV reactor based on a true fluence rate detector. *Environ. Sci. Technol.* 2017, 51, 8094-8100.

[82] Ojeda, C. B.; Rojas, F. S. Process analytical chemistry: applications of ultraviolet/visible spectrometry in environmental analysis: an overview. *Appl. Spectrosc. Rev.* 2009, 44, 245-265.

[83] Adinolfi, V.; Ouellette, O.; et al. Fast and sensitive solution-processed visible-blind perovskite UV photodetectors. *Adv. Mater.* 2016, 28, 7264-7268.

[84] Lin, Q.; Armin, A.; et al. Near infrared photodetectors based on sub-gap absorption in organohalide perovskite single crystals. *Laser Photonics Rev.* 2016, 10, 1047-1053.

[85] Dong, R.; Fang, Y.; et al. High-gain and low-driving-voltage photodetectors based on organolead triiodide perovskites. *Adv. Mater.* 2015, 27, 1912-1918.

[86] Chin, A.; Vaddiraju, S.; et al. Near-infrared semiconductor subwavelength-wire lasers. *Appl. Phys. Lett.* 2006, 88, 163115.

[87] Cheng, S.-Q.; Cai, B.; Zhu, Y.-M. Black silicon as absorber for near-infrared photo-thermal conversion, 2015 Opto-Electronics and Communications Conference (OECC), IEEE: 2015; pp 1-3.

[88] Du, K.-K.; Li, Q.; et al. Control over emissivity of zero-static-power thermal emitters based on phase-changing material GST. *Light Sci. Appl.* 2017, 6, e16194-e16194.

[89] Wang, Z.; Li, Y.; et al. Pure near-infrared to near-infrared up-conversion of multifunctional Tm^{3+} and Yb^{3+} co-doped NaGd $(WO_4)_2$ nanoparticles. *J. Mater. Chem. C* 2014, 2, 4495-4501.

[90] Naczynski, D.; Tan, M.; et al Rare-earth-doped biological composites as in vivo shortwave infrared reporters. *Nature Commun.* 2013, 4, 1-10.

[91] Yakunin, S.; Dirin, D. N.; et al. Detection of gamma photons using solution-grown single crystals of hybrid lead halide perovskites. *Nature Photonics* 2016, 10, 585.

[92] Wei, H.; DeSantis, D.; et al. Dopant compensation in alloyed

$CH_3NH_3PbBr_{3-x}Cl_x$ perovskite single crystals for gamma-ray spectroscopy. *Nature Mater.* 2017, 16, 826-833.

[93] He, Y.; Ke, W.; et al. Resolving the energy of γ-ray photons with $MAPbI_3$ single crystals. *ACS Photonics* 2018, 5, 4132-4138.

[94] Wang, X.; Zhao, D.; et al. PIN diodes array made of perovskite single crystal for X-ray imaging. *Phys. Status Solidi RRL* 2018, 12, 1800380.

[95] Street, R.; Ready, S.; et al. Comparison of PbI_2 and HgI_2 for direct detection active matrix X-ray image sensors. *J. Appl. Phys.* 2002, 91, 3345-3355.

[96] Kabir, M. Z.; Kasap, S. Charge collection and absorption-limited sensitivity of X-ray photoconductors: Applications to *a*-Se and HgI_2. *Appl. Phys. Lett.* 2002, 80, 1664-1666.

[97] Nguyen, V.-C.; Katsuki, H.; et al. Single-crystal perovskite $CH_3NH_3PbBr_3$ prepared by cast capping method for light-emitting diodes. *Jpn. J. Appl. Phys.* 2018, 57, 04FL10.

[98] Chih, Y. K.; Wang, J. C.; et al. NiO_x electrode interlayer and CH_3NH_2/ $CH_3NH_3PbBr_3$ Interface treatment to markedly advance hybrid perovskite-based light-emitting diodes. *Adv. Mater.* 2016, 28, 8687-8694.

[99] Chen, M.; Shan, X.; et al. Manipulating ion migration for highly stable light-emitting diodes with single-crystalline organometal halide perovskite microplatelets. *ACS Nano* 2017, 11, 6312-6318.

[100] Zhang, Q.; Ha, S. T.; et al. Room-temperature near-infrared high-Q perovskite whispering-gallery planar nanolasers. *Nano Lett.* 2014, 14, 5995-6001.

[101] Zhu, H.; Fu, Y.; et al. Lead halide perovskite nanowire lasers with low lasing thresholds and high quality factors. *Nature Mater.* 2015, 14, 636-642.

[102] Liao, Q.; Hu, K.; et al. Perovskite microdisk microlasers self-assembled from solution. *Adv. Mater.* 2015, 27, 3405-3410.

[103] Feng, J.; Yan, X.; et al. "Liquid knife" to fabricate patterning single-crystalline perovskite microplates toward high-performance laser arrays. *Adv. Mater.* 2016, 28, 3732-3741.

[104] Wu, B.; Nguyen, H. T.; et al. Discerning the surface and bulk recombination kinetics of organic–inorganic halide perovskite single crystals. *Adv. Energy Mater.* 2016, 6, 1600551.

[105] Murali, B.; Yengel, E.; et al. The surface of hybrid perovskite crystals: a boon or bane. *ACS Energy Lett.* 2017, 2, 846-856.

[106] Yang, Y.; Yang, M.; et al. Top and bottom surfaces limit carrier lifetime in lead iodide perovskite films. *Nature Energy* 2017, 2, 1-7.

[107] Yuan, Y.; Huang, J. Ion migration in organometal trihalide perovskite and its impact on photovoltaic efficiency and stability. *Acc. Chem. Res.* 2016, 49, 286-293.

[108] Deng, Y.; Xiao, Z.; Huang, J. Light-induced self-poling effect on organometal trihalide perovskite solar cells for increased device efficiency and stability. *Adv. Energy Mater.* 2015, 5, 1500721.

[109] Li, W.; Fan, J.; et al. Aquointermediate assisted highly orientated perovskite thin films toward thermally stable and efficient solar cells. *Adv. Energy Mater.* 2017, 7, 1601433.

[110] Grancini, G.; D'Innocenzo, V.; et al. $CH_3NH_3PbI_3$ perovskite single crystals: surface photophysics and their

interaction with the environment. *Chem. Sci.* 2015, 6, 7305-7310.

[111] Ecker, B. R.; Wang, C.; et al. Intrinsic behavior of $CH_3NH_3PbBr_3$ single crystals under light illumination. *Adv. Mater. Interfaces* 2018, 5, 1801206.

[112] Williams, A. C.; Barry, B. W. Penetration enhancers. *Adv. Drug Deliv. Rev.* 2012, 64, 128-137.

[113] Lyu, M.; Yun, J.-H.; et al. Organic–inorganic bismuth (III)-based material: A lead-free, air-stable and solution-processable light-absorber beyond organolead perovskites. *Nano Res.* 2016, 9, 692-702.

[114] Leng, M.; Chen, Z.; et al. Lead-free, blue emitting bismuth halide perovskite quantum dots. *Angew. Chem. Int. Ed.* 2016, 55, 15012-15016.

Chapter 7

Optical Properties of Single Crystals

Senthilkumar Chandran and Srinivasan Manikam

Abstract

Nonlinear optical crystals plays important role in the field of photo electronics, optical communication, optical modulators, laser spectroscopy, frequency conversion and so on. Semi-organic crystals exhibit high NLO response, thermal stability, laser damage threshold, mechanical stability, wide optical window transmittance and structural diversity. Combinations of inorganic and organic molecules yield the semi-organic crystals. Based on its structural diversity it's classified into three categories. In this chapter explains various kinds of semi-organic crystals and their optical, thermal, mechanical, laser damage threshold value and NLO properties and also explains the importance of these crystals in the field of optoelectronics, frequency conversion and other optical applications.

Keywords: nonlinear optical materials, optoelectronics, semi-organic single crystals, laser damage thresholds, optical device

1. Introduction

Over the last three decades, the discovery of new crystals for optical applications has been an emerging area of research. Nonlinear optical crystals are significant in science and modern technology because of their technological importance in the areas of optical communication, optical modulators, laser spectroscopy, frequency conversion, optical bi-stable devices, electro-optical device applications in photonics technology, optoelectronics, information processing, sensors, laser technology, frequency doubling and color displays. Optical applications depend upon various physical features, such as refractive index, birefringence, thermal stability and physicochemical behaviors. Materials with high second-order optical nonlinearity, high optical transmittance with low cut-off wavelength, high laser damage threshold value and easy growth with large dimensions are needed to understand many of these applications. The growth of the new kind of optical crystals with good physical and chemical properties are very important in optoelectronics, photonics laser processing and other applications. The search for high non-linear optical crystals for efficient signal processing has been stimulated by optoelectronics [1–8].

In the field of optoelectronics and photonics, nonlinear optical (NLO) materials are capable of generating the second harmonic frequency. In various device applications, nonlinear optical (NLO) crystals with high conversion efficiency for the second harmonic generation (SHG) and transparent in ultraviolet–visible regions are required. In the current technology world, there is a lot of competition for powerful nonlinear optical devices to satisfy the day-to-day requirements. Usually, organic materials show excellent nonlinear optical (NLO) characteristics. Due to this reason, it becomes important to grow a more and more new organic-based single crystal. Most of the scientists have been focused their research on organic compounds over the past decades as it shows high nonlinear coefficients compared to inorganic materials. But apart from their nonlinearity, the organic molecules are attached with weak van der Walls and hydrogen bonds with π conjugated electrons that make the organic materials are soft, poor physico-chemical stability, low mechanical strength and difficult to polish. Further, these materials have strong absorption in the UV region. On the other hand, the inorganic materials have high laser damage threshold, high melting point and high mechanical strength, but these materials possess moderate NLO behavior. Compare to organic and inorganic materials, the semi-organic materials show combining the properties of both organic and inorganic materials. In this view, semi-organic materials must be analyzed [4–6, 9–12].

2. Selection criteria for NLO materials

In reality, there is no possibility to obtain the perfect nonlinear crystal. The applicability of a specific crystal depends on the nonlinear optical method used, the desired device features and the pump laser. In one application, unique and important material properties may not be relevant in another application. For example, a material with a large angular bandwidth requires the efficient doubling of very high power lasers with poor beam quality. A crystal, which has a lower nonlinearity but permits noncritical phase match angle, will work better than one, which would be more nonlinear, but is critically phase-matched. On the other side, one with a large nonlinearity would be the ideal material for the doubling of femtosecond light signal, even if a very thin crystal can be used to prevent the dispersive expansion of the second harmonic output signals. Nonlinear frequency converters are frequently employed with a capable non-tunable laser source. Kurtz and Perry powder SHG technique was introduced in 1960. In this method, the fine powdered material is irradiated with laser and scattered light is collected and studied for its harmonic capacity with the help of proper filters. This is a rapid and qualitative analyzing method for second-order NLO effect and this technique is suitable for inorganic, organic, semi-organic and new materials. To understand the nonlinear optical effect, the appropriate medium is essential. A non-centrosymmetric crystal, which shows the following characteristics, is required for nonlinear device fabrications (**Table 1**):

1. Good optical quality with large dimensions

2. Wide transmittance with low cut-off wavelength

3. High thermal-mechanical and chemical stability

4. High laser damage threshold value

5. Large birefringence

6. Low absorption cut-off wavelength

7. High second-order nonlinear optical coefficient

8. Easy to device fabrication

S.no	Laser conditions	Crystal parameters
1	Environment	Temperature, Moisture acceptance
2	Beam size	Crystal dimension, spatial walk-off
3	Bandwidth	Spectral acceptance
4	Divergence	Acceptance angle
5	NLO method	Kind of phase- Matching
6	Repetition rate	Surface damage threshold

Table 1.
Parameters for selecting a nonlinear optical crystal.

3. NLO crystal of current interest

Significant developments in new technologies are responsible for the development of new crystals of superior quality. The high-speed and significant amount of optical parallelism would ultimately lead to optoelectronics devices which were a wide range of optical functions will be implemented. The growth of photonic technology, however, depends mainly on the progress made in developing new optical material with improved performance. Crystals with the nonlinear optical response (NLO) are expected to play an important role in facilitating optoelectronic and photonic developments. In optical and electro-optical applications, several NLO crystals have been found out as potential candidates. With the development of many devices using solid-state laser sources, nonlinear optical crystals have received special revolution. For the manufacture of electro-optic modulators that converts an electrical signal into an optical signal and transmission on a fiber optic cable, NLO crystals are very important. Currently, such devices are made with inorganic NLO materials. In this view, further new other crystals should be developed. Recently researcher is showing their keen interest in the development of organic–inorganic salts which show good optoelectronics characteristics [5–12].

4. Semi organic single crystals

Hybrid inorganic–organic structure materials constitute a crucial class of materials that have been extensively analyzed in the last few decades owing to their possible applications in the dielectric, optical luminescence, magnetic, and electronic properties. Because of their tremendous chemical and structural diversity and many technologically applicable properties, these are the fastest-growing fields in materials science. Hybrid inorganic–organic structural materials characterize new generations of crystalline solid-state materials which are formed from metal ions and organic linkers. It is suggested that the semi-organic crystals would have the properties of both inorganic and organic materials. There are three types of semi-organic crystals exist [1–3, 8]:

1. Organic–Inorganic salts

2. Metal–Organic coordination complexes

3. Organometallic compounds

4.1 Organic–inorganic salts

Organic–inorganic salt L-arginine phosphate monohydrate (LAP) was explored in 1971. LAP is a promising biaxial nonlinear optical (NLO) crystal. It crystallizes in the monoclinic crystal structure with P_{21} space group. The chemical formula of LAP is $C_6H_{14}N_4O_2H_3PO_4.H_2O$ [13, 14]. In the past two decades, the growth of LAP crystal and its NLO, electrical, mechanical, optical, surface, thermal and other properties have been well investigated. It has a large effective NLO coefficient. The SHG efficiency is 3.5 times higher than the standard potassium dihydrogen phosphate (KDP) crystal. It has high surface Laser damage threshold (higher than 1GW/ cm^2 at 1064 nm), high nonlinear coefficient (>1 pm/V), wide transmission range (220 nm–1950 nm), low hygroscopicity nature and high-frequency conversion efficiency (38.9%). Compared to KDP, it is more sensible and replaces the need for KDP for laser fusion experiments in harmonic frequency generation. The high laser damage threshold of the LAP crystal indicates the advantage of high-power laser devices in frequency conversion and the critical benefit of modern photonic devices. LAP is easy to grow by solution growth technique into a large size single crystal and has improved physicochemical stability than the standard KDP crystal. LAP belongs to KDP family but is particularly hybridized by an amino acid. From the perception of structure, it contains the alternate layers of the acentric tetrahedral inorganic dihydrogen phosphate anionic groups (H_2PO_4), the water molecules (H_2O) and the organic chiral L-arginine molecule $[(H_2N)CNH(CH_2)_3CH(NH_3)^-COO^+]$, combined by positive–negative coulomb interactions and hydrogen bonds (**Figure 1**). Planar guanidinium and carboxylate groups are thought to be responsible for the SHG at the two ends of L-arginine. In many aspects, the LAP crystal can match with the inorganic NLO potassium dihydrogen phosphate crystal. Furthermore, it has a larger NLO coefficient d_{21} = 2.14 d_{36}(KDP). It has been analyzed it is one of the good material for ultrafast signal and powerful laser doubling techniques [8, 13–17].

Considerable research has been made to the growth of LAP crystal derivatives, with the expectation of the enactment of NLO performance, surface damage threshold, mechanical strength and optical transmittance. The deuterated LAP (DLAP) is one of its analogues with improved IR transmittance and a wider frequency conversion region. DLAP was assumed to be an efficient material to replace the conversion efficiency of the KDP crystal. DLAP can be used in laser fusion experiment and many high-tech areas. DLAP is the deuterated form of LAP, in which the protons have been replaced. The absorption noted around 1000 nm, in

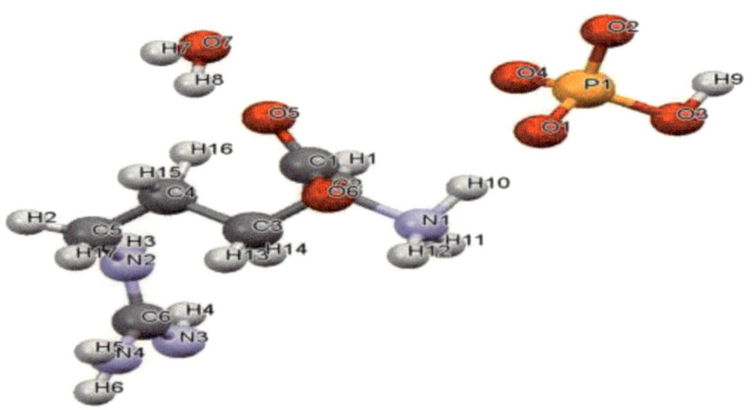

Figure 1.
Molecular structure of L-arginine phosphate monohydrate.

LAP owing to the overtones of vibration linked with hydrogen-containing groups, which is efficiently reduced in DLAP. DLAP also has a higher surface damage threshold than the LAP for Nd: YAG laser light [18].

4.2 Metal–organic coordination complexes

These forms of semi-organic crystals are studied by several polyhedral with a central metal ion coordinated by several organic and inorganic ligands. The common formula of metal–organic coordination complexes is MM' LnLm', where MM'- the various metal ions and LnLm' the organic and/or inorganic ligands. The organic ligand (L) is generally more dominant in the nonlinear optical effect. The II (B) divalent metal (Zn, Cd, Hg) complexes have high transparency in the ultra-violet region because of their locked d^{10} shell. The metal–organic complexes can be categorized into three groups [1, 2, 8]:

1. Island type

2. Chain type

3. Network type

4.2.1 Island type

In this type-specific coordination, polyhedra are held together only by comparatively weak intermolecular interactions (van der Waals forces, hydrogen bonding, long-distance coulomb interactions). Several thiourea complexes were synthesized and tested for their powder efficiencies based on the intuitive approach of incorporating asymmetric conjugated organic molecules into inorganic distorted polyhedra. Bisthiourea cadmium chloride, triallylthiourea cadmium chloride, triallylthiourea cadmium bromide, triallylthiourea mercury chloride and zinc tris(thiourea) sulfate were recognized as the efficient NLO materials. These are some important island type metal–organic coordination complexes. Metal complexes of thiourea have low UV cut -off wavelengths, therefore these materials are suitable for high power frequency conversion applications. These materials can be used as better alternatives for KDP crystals in frequency doubling and laser fusion experiments due to their higher values of laser damage threshold.

Bis(thiourea) cadmium chloride (BTCC) is an efficient NLO material for SHG applications. The chemical formula of BTCC is ($Cd[SC(NH_2)_2]Cl_2$) and belongs to the orthorhombic crystal class with the space group Pmn_{21}. The SHG efficiency of BTCC is almost the same as that of urea. Compared to the other solution-grown NLO crystals, the BTCC crystal has a greater laser damage threshold value. The single-shot damage thresholds resistance of BTCC is 32 GW/cm^2 at 532 nm. The transmission range varies from 285 nm to 1900 nm. Considerable absorption observed around 1500 nm. It has good transmission in the entire visible range. Remarkable transmission in the spectrum is due to the reduction in absorption. This is owing to the number of N-H bonds, which reasons absorption around 1040 nm by vibrational overtones is slighter in BTCC than in ZTS. It also has good mechanical behavior which is comparable with KDP crystal. The NLO coefficient of BTCC is $d_{11} = 2.75d_{36}$(KDP). Along the phase-matching angle, its SHG efficiency is almost similar to that of urea. It is a promising NLO crystal for different applications in the area of laser and optoelectronic [8, 19–21].

Another important island type metal–organic coordination complex is zinc tris(thiourea) sulfate (ZTS). The large size of ZTS crystal can be easily grown using an aqueous solution. The chemical formula of ZTS is $Zn[CS(NH_2)_2]_3SO_4$. It also belongs to the orthorhombic crystal class with space group Pca_{21}. It is a good material for nonlinear engineering applications which has high nonlinearity. It has good optical quality, with low defect densities. The single-shot damage threshold of ZTS crystal is found to be 40 GW/cm^2. BTCC has only two thiourea units while ZTS own three. This decrease in absorption at around 1064 nm which causes significant contribution towards an enhancement in the laser damage resistance of the ZTS crystal. The ZTS has lower cut-off below 300 nm, which is valuable in semi-organic NLO crystals over their organic crystals. It has allowed angular sensitivity that shows useful for type-II second-harmonic generation. It has nearly 1.2 times higher nonlinear than KDP crystal. High surface damage threshold and good transmittance make it a better alternative crystal for KDP crystal in laser fusion and frequency-doubling experiments [22–24].

4.2.2 Chain type

In this kind coordinate polyhedra are connected through chemical bonds, corner by corner or edge by edge, coordinating one-dimensional polymers in the crystal structure. Thiosemicarbazide cadmium chloride monohydrate (TSCCC) and, thiosemicarbazide cadmium bromide monohydrate (TSCCB) are examples of chain type NLO crystals. The chemical formula of TSCCC is $Cd(NH_2CSNHNH_2)Cl_2.H_2O$. Good quality of TSCCC crystal can be easily grown using slow evaporation technique. It crystallizes in the monoclinic system with non-centrosymmetric space group Cc. The SHG efficiency of TSCCC is 14 times more than KDP crystal, this may be due to the chlorine atom must be affected in the coordinate polyhedral. TSCCC crystal has good transmittance in the UV–NIR region and cut-off wavelength is below 280 nm, which is enough for SHG laser radiation of 1064 nm and other optical applications. The laser damage threshold energy value is calculated to be 725 MW/cm at wavelength of 1064 nm. The thermal stability of TSCCC is 148°c which is higher than that of LAP crystal (144°c). It also has good mechanical strength, high dielectric constant and low dielectric losses. TSCC crystals show negative photoconductivity which can be used for IR detector applications. The third-order nonlinear susceptibility of the crystal is calculated to be 2.774×10^{-5} esu which shows it as a suitable material for nonlinear optical applications [8, 25–27].

Another important chain form of NLO material is thiosemicarbazide cadmium bromide monohydrate (TSCCB). The chemical formula of TSCCB (**Figure 2**) is $Cd(NH_2(SNHNH_2)Br_2.H2O$. TSCCB belong to the Cc space group with similar cell parameters and similar molecular packing style of TSCCB. The lower cut-off

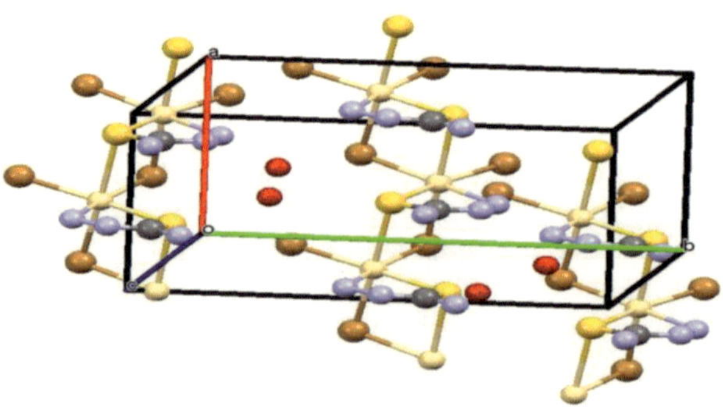

Figure 2.
Packing diagram of thiosemicarbazide cadmium bromide monohydrate.

wavelength is calculated to be 293 nm. The crystal has a wide transparency window in the entire visible region, which confirms the suitability of the TSCCB in NLO device applications. The crystal has high mechanical stability, high thermal stability (190°c), high dielectric constant and low dielectric losses. The SHG efficiency is calculated to be 1.98 times that of standard KDP crystal. The non-linearity of the TSCCB occurs due to the organic–inorganic ring structure of the complex material containing the cadmium ions [8, 28–32].

4.2.3 Network type

In this form, two or three-dimensional coordinative chemical bonds join all polyhedral composed. This is the most referring type because its physicochemical stability and nonlinear optical behavior can be greatly increased compared to organic crystals. The $MHg(SCN)_4$ series complex has three-dimensional networks. This kind of material possesses good NLO properties, which are relevant to their structural features [1–3, 8, 33, 34].

4.3 Organometallic compounds

In the late 1960s, organometallic crystals were brought into use for optoelectronic applications to resolve the physicochemical instability of organic crystals. Based on the concept of combining the inorganic distorted polyhedron with an asymmetric conjugate organic molecule, organometallic compounds can be synthesized. In optoelectronics and nonlinear optical fields, organometallic crystals are of great interest because such crystals have the potential to combine the organics, high optical nonlinearity and chemical flexibility with temporal, thermal stability and strong inorganic transmittance. The high resistance to surface laser damage threshold is another important advantage of organometallic materials. To understand the use of organometallic substances for device applications, the growth of organometallic single crystals has been subject to perennial concern. Like organic molecules, organometallic compounds also offer the advantages of architectural versatility and ease of processing and tailoring.

Recent analysis shows that the organometallic complexes, in particular, the metallic complexes of thiocyanate and their Lewis base have impressed the material scientist because of their strong second-order nonlinear optical efficiency and stable physio-chemical properties. The molecular engineering of bimetallic complexes of thiocyanate crystallizes into non-centrosymmetric space group which affirms the second harmonic generation and also enabled the crystal growers to grow the novel varieties of NLO crystals. Considerable nonlinearity occurs in organometallic thiocyanate crystals because of the delocalized π-electrons from ligand to metal or metal to ligand. Increased NLO properties also occur in the diversity of central metal atoms, oxidation levels, size and nature of the ligand. These materials also have high nonlinearity, high surface damage threshold, low UV cut-off wavelength, moderate thermal and mechanical properties. Crystals such as manganese mercury thiocyanate (MMTC), cadmium mercury thiocyanate (CMTC), zinc mercury thiocyanate (ZMTC), manganese mercury thiocyanate bis dimethyl sulphoxide (MMTD), and zinc cadmium thiocyanate (ZCTC) belong to this type. Most of the organometallic thiocyanate crystals discussed here were grown by solution growth technique (slow evaporation method, slow cooling method and temperature lowering method). Normally, organometallic thiocyanate crystals can be classified into manganese mercury thiocyanate (MMTC), cadmium mercury thiocyanate (CMTC), zinc mercury thiocyanate (ZMTC) and so on [33–45].

4.3.1 Manganese mercury thiyocyanate

The manganese mercury thiyocyanate ($MHg(SCN)_4$) series of crystalline complexes have been known for a century in analytical chemistry for their characteristic shapes and colors. Although SCN ion is a good chromophore for second-order NLO properties, however, these structure type crystals are found to be crystallized in centrosymmetric space groups, which leads them to lose their macroscopic NLO properties or crystallize in a noncentrosymmetric space group, unfortunately, the low energy d-d transitions present in these compounds, normally observed in the visible light region and limit their NLO usefulness. MMTC crystal belongs to the tetragonal crystallographic system with space group $I\bar{4}$. The SHG efficiency of MMTC is 18 times that of urea. Thus, the second harmonic generation efficiency of MMTC is very much higher than that of other organometallic family crystals such as CMTC, CMTD and BTCC and other laser materials like KDP, LAP and BBO. The laser damage threshold of MMTC was found to be 10.5 GW/cm2, which suggests that the MMTC has high damage threshold value than the KDP and BBO. The crystal is thermally stable up to 353°C. MMTC has a large transmission window range from 373 nm to 2250 nm without any absorption peak. The UV cut-off wavelength of MMTC is 383 nm, which is nearly the same as that of CMTC. Third-order nonlinear susceptibility ($\chi^{(3)}$) of MMTC is found to be 3.13×10^{-8} esu. The hardness properties of MMTC (50 kg/mm2) is more than CMTD (47 kg/mm^2) and less than ZTS (116 kg/mm^2) and BTCC (136 kg/mm^2). The high SHG efficiency, wide optical transmittance, high thermal stability and moderate mechanical property of MMTC show that this crystal is an excellent material for photonic device fabrication [33–37].

4.3.2 Cadmium mercury thiocyanate

Cadmium mercury thiocyanate and zinc mercury thiocyanate and are well known efficient phase matchable second harmonic generation single crystals [39]. Bimetallic thiocyanates are semi-organic compounds, with chemical formula AB(SCN)4, that exhibit high nonlinearity. Among the bimetallic thiocyanate materials, zinc mercury thiocyanate (ZMTC) and cadmium mercury thiocyanate (CMTC) is found to have all the important characteristics such as crystallization in a noncentrosymmetric space group, colorless, and high thermal stability. Both ZMTC and CMTC are SHG crystals which can convert 1064 nm radiation. Cadmium mercury thiocyanate ($CdHg(SCN)_4$) is widely studied organometallic crystals. It belongs to the tetragonal system with space group $I\bar{4}$. The crystal is thermally stable up to 251°C. UV–Vis cut-off wavelength of CMTC was found to be 383 nm which indicates the potential of generating blue-violet light using a diode laser. The laser damage threshold value of CMTC is calculated to be 11.14 MW/cm^2. The optical limiting threshold value of the CMTC is calculated to be 31.3 mW, which shows an excellent optical limiting property of CMTC. The phase-matching angles are $\theta = 47.7°$ and $\phi = 0^{\circ}$, Third-order nonlinear susceptibility ($\chi^{(3)}$) calculated to be 14.27×10^{-6} (esu). The SHG efficiency of CMTC is 11.3 times higher than that of Urea. The results show that this crystal has high NLO coefficient and very high laser damage threshold value which confirms this crystal can be used for many optoelectronic device applications including high power frequency conversion and fabrication of optical limiting devices [8, 33, 40–42].

4.3.3 Zinc mercury thiocyanate

Zinc mercury thiocyanate ($ZnHg(SCN)_4$) is well known NLO material for SHG of 1064 nm radiation. ZMTC (**Figure 3**) belongs to the tetragonal crystal system.

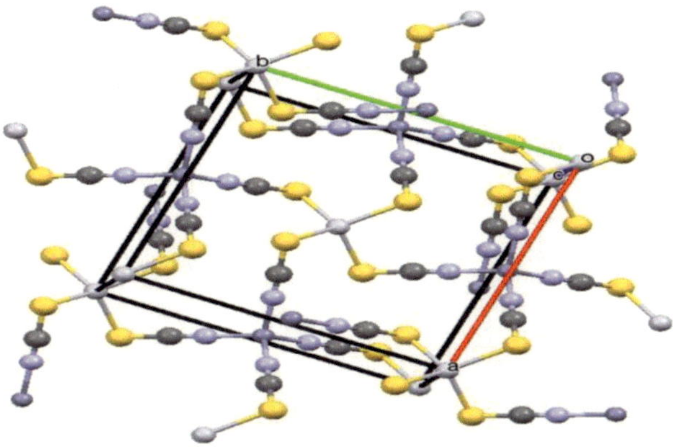

Figure 3.
Three dimensional views of zinc mercury thiocyanate.

UV transparency cut-off wavelength is 257 nm. It is thermally stable up to 310°C. The second harmonic generation efficiency is found to be 14 times greater than that of urea [33, 43, 45].

5. Conclusion

Combination of inorganic and organic material is a potential approach to generate more stable NLO crystals. The chemistry correlated with this group of materials is such that the designing of new systems is extremely versatile and potentially contributing to improved optical response. Various semi-organic crystal satisfies the basic criteria's of NLO applications. The remarkable properties of these crystals suggest that this crystal can be used in various optical device applications.

Acknowledgements

The authors are grateful thank to CCDC for accessing the crystal structures.

Author details

Senthilkumar Chandran[1*] and Srinivasan Manikam[2]

1 Department of Physics, Government Arts and Science College, Hosur, Tamilnadu, India

2 Department of Physics, Centre for Crystal Growth, SSN College of Engineering, Kalavakkam, Tamilnadu, India

*Address all correspondence to: senthilkumarchandran89@gmail.com

References

[1] Braga D, Grepioni F, Desiraju GR. Crystal engineering and organometallic architecture. Chemical reviews. 1998;98(4):1375-1406. DOI: 10.1021/cr960091b

[2] Long NJ. Organometallic Compounds for Nonlinear Optics—The Search for En-light-enment!. Angewandte Chemie International Edition in English. 1995 ;34(1):21-38. DOI: 10.1002/anie.199500211

[3] Tan JC, Cheetham AK. Mechanical properties of hybrid inorganic–organic framework materials: establishing fundamental structure–property relationships. Chemical Society Reviews. 2011;40(2):1059-1080. DOI: 10.1039/C0CS00163E

[4] Ramajothi J, Dhanuskodi S, Nagarajan K. Crystal growth, thermal, optical and microhardness studies of tris (thiourea) zinc sulphate-a semiorganic NLO material. Crystal Research and Technology: Journal of Experimental and Industrial Crystallography. 2004;39(5):414-20. DOI: 10.1002/crat.200310204

[5] Aggarwal MD., Stephens J, Batra AK, Lal RB. Bulk crystal growth and characterization of semiorganic nonlinear optical materials. Journal of Optoelectronics and Advanced Materials. 2003; 5 (3): 555-562.

[6] Chandran S, Paulraj R, Ramasamy P. Nucleation kinetics, crystal growth and optical studies on lithium hydrogen oxalate monohydrate single crystal. Journal of Crystal Growth. 2017 ;468:68-72.DOI: 10.1016/j.jcrysgro.2016.11.006

[7] Chandran S, James GJ, Magesh M, Prasanna N. Synthesis, crystal growth, structural, spectral, laser threshold energy and dielectric properties of lithium L-tartrate monohydrate crystal. Journal of Molecular Structure.2020;1223:128988. DOI: 10.1016/j.molstruc.2020.128988

[8] Jiang MH, Fang Q. Organic and semiorganic nonlinear optical materials. Advanced Materials. 1999 ;11(13):1147-51. DOI: 10.1002/(SICI)1521-4095(199909)11:13<1147::AID-ADMA1147>3.0.CO;2-H

[9] Siddheswaran R, Sankar R, Rathnakumari M, Murugakoothan P, Jayavel R, Sureshkumar P. Growth and characterization of a new semi-organic non-linear optical crystal l-arginine hydrochlorofluoride monohydrate (lahclf). Surface Review and Letters. 2006;13(06):803-808. DOI: 10.1142/S0218625X06008888

[10] Zaccaro J, Salvestrini JP, Ibanez A, Ney P, Fontana MD. Electric-field frequency dependence of Pockels coefficients in 2-amino-5-nitropyridium dihydrogen phosphate organic–inorganic crystals. JOSA B. 2000; 17(3):427-432. DOI: 10.1364/JOSAB.17.000427

[11] Suresh S, Ramanand A, Jayaraman D, Mani P. Review on theoretical aspect of nonlinear optics. Rev. Adv. Mater. Sci. 2012;30(2):175-183.

[12] Thukral K, Vijayan N, Haranath D, Maurya KK, Philip J, Jayaramakrishnan V. Comprehensive study on l-Proline Lithium Chloride Monohydrate single crystal: A semiorganic material for nonlinear optical applications. Arabian Journal of Chemistry. 2019;12(8):3193-3201.DOI: 10.1016/j.arabjc.2015.08.022

[13] Aoki K, Nagano K, Iitaka Y. The crystal structure of L-arginine phosphate monohydrate. Acta Crystallographica Section B: Structural Crystallography and Crystal Chemistry. 1971 ;27(1):11-23. DOI: 10.1107/s056774087100164x

[14] Hameed AH, Ravi G, Ilangovan R, Azariah AN, Ramasamy P. Growth and characterization of deuterated analog of l-arginine phosphate single crystals. Journal of crystal growth. 2002;237:890-893. DOI: 10.1016/S0022-0248(01)02062-0

[15] Wankhade PM, Muley GG. Study on effect of 1, 3-dimethyl urea doping on optical properties of l-arginine phosphate monohydrate (LAP) single crystal. Results in Physics. 2013;3:97-102. DOI: 10.1016/j.rinp.2013.06.002

[16] Peramaiyan G, Pandi P, Bhagavannarayana G, Kumar RM. Studies on growth, structural, dielectric, laser damage threshold, linear and nonlinear optical properties of methylene blue admixtured L-arginine phosphate single crystal. Spectrochimica Acta Part A: Molecular and Biomolecular Spectroscopy. 2012 Dec 15;99:27-32. DOI: 10.1016/j.saa.2012.08.087

[17] Yokotani A, Sasaki T, Fujioka K, Nakai S, Yamanaka C. Growth and characterization of deuterated L-arginine phosphate monohydrate, a new nonlinear crystal, for efficient harmonic generation of fusion experiment lasers. Journal of Crystal Growth. 1990 Jan 1;99(1-4):815-819. DOI: 10.1016/S0022-0248(08)80032-2

[18] Yokotani A, Sasaki T, Yoshida K, Nakai S. Extremely high damage threshold of a new nonlinear crystal L-arginine phosphate and its deuterium compound. Applied physics letters. 1989 Dec 25;55(26):2692-2693. DOI: 10.1063/1.101969

[19] Ushasree PM, Muralidharan R, Jayavel R, Ramasamy P. Growth of bis (thiourea) cadmium chloride single crystals—a potential NLO material of organometallic complex. Journal of crystal growth. 2000 ;218(2-4):365-371. DOI: 10.1016/S0022-0248(00)00593-5

[20] Venkataramanan V, Maheswaran S, Sherwood JN, Bhat HL. Crystal growth and physical characterization of the semiorganic bis (thiourea) cadmium chloride. Journal of crystal growth. 1997 Aug 2;179(3-4):605-610. DOI: 10.1016/S0022-0248(97)00137-1

[21] Dhumane NR, Hussaini SS, Dongre VG, Karmuse PP, Shirsat MD. Growth and characterization of glycine doped bis thiourea cadmium chloride single crystal. Crystal Research and Technology: Journal of Experimental and Industrial Crystallography. 2009;44(3):269-274. DOI: 10.1002/crat.200800239

[22] Venkataramanan V, Subramanian CK, Bhat HL. Laser induced damage in zinc tris (thiourea) sulfate and bis (thiourea) cadmium chloride. Journal of applied physics. 1995 Jun 1;77(11):6049-6051. DOI: 10.1063/1.359127

[23] Ushasree PM, Muralidharan R, Jayavel R, Ramasamy P. Metastable zonewidth, induction period and interfacial energy of zinc tris (thiourea) sulfate. Journal of crystal growth. 2000 Mar 1;210(4):741-745. DOI: 10.1016/S0022-0248(99)00900-8

[24] Verma S, Singh MK, Wadhawan VK, Suresh CH. Growth morphology of zinc tris (thiourea) sulphate crystals. Pramana. 2000 Jun 1;54(6):879-888. DOI: 10.1007/s12043-000-0183-1

[25] Sankar R, Raghavan CM, Kumar RM, Jayavel R. Growth and characterization of a new semiorganic non-linear optical thiosemicarbazide cadmium chloride monohydrate (Cd (NH2NHCSNH2) Cl2· H2O) single crystals. Journal of crystal growth. 2007 ;305(1):156-161. DOI: 10.1016/j.jcrysgro.2007.03.019

[26] Maadeswaran P, Thirumalairajan S, Chandrasekaran J. Growth, thermal, optical and birefringence studies

of semiorganic nonlinear optical thiosemicarbazide cadmium chloride monohydrate (TCCM) single crystals. Optik. 2010;121(9):773-777. DOI: 10.1016/j.ijleo.2008.09.041

[27] Akilan M, Ragu R, Angelena JP, Das SJ. Investigation on nucleation kinetics, growth, optical, mechanical, conductivity and Z-scan studies on thiosemicarbazide cadmium chloride monohydrate (TSCCCM) single crystals for nonlinear applications. Journal of Materials Science: Materials in Electronics. 2019 ;30(16):15116-15129. DOI: 10.1007/s10854-019-01884-y

[28] Nicolo F, El Ghaziri HA, Chapuis G. Structure of dibromo (thiosemicarbazide) cadmium (II) monohydrate. Acta Crystallographica Section C: Crystal Structure Communications. 1988;44(6):975-977. DOI: 10.1107/S0108270188000885

[29] Chandrasekaran J, Ilayabarathi P, Maadeswaran P. Spectroscopic, thermal, optical, dielectric and mechanical properties of thiosemicarbazide cadmium bromide (tsccb): a semiorganic nlo crystal. 2011; 4:431-436

[30] Maadeswaran P, Thirumalairajan S, Karthika P, Chandrasekaran J. Growth and characterization of a semiorganic nonlinear optical crystal-Cadmium thiosemicarbazide bromide. Journal of Optoelectronics and Biomedical Materials Vol. 2009;1:180-187.

[31] Prakash JT, Gnanaraj JM. Growth and characterization of Cadmium Thiosemicarbazide Bromide crystals for antibacterial and nonlinear optical applications. Spectrochimica Acta Part A: Molecular and Biomolecular Spectroscopy. 2015 ;135:25-30. DOI: 10.1016/j.saa.2014.06.120

[32] Wang WS, Sutter K, Bosshard C, Pan Z, Arend H, Günter P, Chapuis G, Nicolo F. Optical second-harmonic generation in single crystals of thiosemicarbazide cadmium bromide hydrate (Cd (NH2NHCSNH2) Br2· H2O). Japanese journal of applied physics. 1988;27(7R):1138. DOI: 10.1143/JJAP.27.1138

[33] Hegde TA, Dutta A, Gandhiraj V. Review on growth and characterization of nonlinear optical organometallic thiocyanate crystals. International Journal of Engineering and Technology Innovation. 2019 ;9(4):257.

[34] Wang XQ, Xu D, Lu MK, Yuan DR, Xu SX. Crystal growth and characterization of the organometallic nonlinear optical crystal: manganese mercury thiocyanate (MMTC). Materials research bulletin. 2001;36(5-6):879-887. DOI: 10.1016/ S0025-5408(01)00573-6

[35] Wang XQ, Xu D, Lu MK, Yuan DR, Xu SX, Guo SY, Zhang GH, Liu JR. Crystal growth and characterization of a novel organometallic nonlinear-optical crystal:: MnHg (SCN) 4 (C2H6OS) 2. Journal of crystal growth. 2001 ;224(3-4):284-93. DOI: 10.1016/ S0022-0248(01)01012-0

[36] Joseph GP, Philip J, Rajarajan K, Rajasekar SA, Pragasam AJ, Thamizharasan K, Kumar SR, Sagayaraj P. Growth and characterization of an organometallic nonlinear optical crystal of manganese mercury thiocyanate (MMTC). Journal of crystal growth. 2006;296(1):51-57. DOI: 10.1016/j.jcrysgro.2006.08.023

[37] Usha RJ, Sagayaraj P, Joseph V. Linear and nonlinear optical, mechanical, electrical and surface studies of a novel nonlinear optical crystal–Manganese mercury thiocyanate (MMTC). Spectrochimica Acta Part A: Molecular and Biomolecular Spectroscopy. 2014;133:241-249. DOI: 10.1016/j.saa.2014.04.161

[38] Pearson RG. Hard and soft acids and bases. Journal of the American

Chemical society. 1963 ;85:3533-3539. DOI: 10.1021/ja00905a001

[39] Sturmer W, Deserno U. Mercury-thiocyanate-complexes: Efficient phase-matchable optical SHG in crystal class 10. Physics Letters A. 1970;32(7):539-540. DOI: 10.1016/0375-9601(70)90497-4

[40] Yuan D, Zhong Z, Liu M, Xu D, Fang Q, Bing Y, Sun S, Jiang M. Growth of cadmium mercury thiocyanate single crystal for laser diode frequency doubling. Journal of crystal growth. 1998;186(1-2):240-244. DOI: 10.1016/S0022-0248(97)00461-2

[41] Hegde TA, Dutta A, Vinitha G. $\chi^{(3)}$ measurement and optical limiting behaviour of novel semi-organic cadmium mercury thiocyanate crystal by Z-scan technique. Applied Physics A. 2018;124(12):808. DOI: 10.1007/s00339-018-2235-8

[42] Yuan D, Xu D, Liu M, Qi F, Yu W, Hou W, Bing Y, Sun S, Jiang M. Structure and properties of a complex crystal for laser diode frequency doubling: Cadmium mercury thiocyanate. Applied physics letters. 1997;70(5):544-546. DOI: 10.1063/1.118335

[43] Xu D, Yu WT, Wang XQ, Yuan DR, Lu MK, Yang P, Guo SY, Meng FQ, Jiang MH. Zinc mercury thiocyanate (ZMTC). Acta Crystallographica Section C: Crystal StructureCommunications. 1999;55(8):1203-5.DOI: 0.1107/S0108270199005983

[44] Kumari PN, Kalainathan S, Raj NA. Study of optimum growth condition and characterization of zinc mercury thiocyanate (ZMTC) single crystals in silica gel. Materials Research Bulletin. 2007;42(12):2099-2106.DOI: /10.1016/j.materresbull.2007.01.018

[45] Wang XQ, Xu D, Yuan DR, Tian YP, Yu WT, Sun SY, Yang ZH, Fang Q, Lu MK, Yan YX, Meng FQ. Synthesis, structure and properties of a new nonlinear optical material: zinc cadmium tetrathiocyanate. Materials Research Bulletin. 1999 ; 34(12-13):2003-2011. DOI: 10.1016/S0025-5408(99)00211-1

Chapter 8

Coordination Polymer Frameworks for Next Generation Optoelectronic Devices

Hemali Rathnayake and Sheeba Dawood

Abstract

Metal–organic frameworks (MOFs), which belong to a sub-class of coordination polymers, have been significantly studied in the fields of gas storage and separation over the last two decades. There are 80,000 synthetically known MOFs in the current database with known crystal structures and some physical properties. However, recently, numerous functional MOFs have been exploited to use in the optoelectronic field owing to some unique properties of MOFs with enhanced luminescence, electrical, and chemical stability. This book chapter provides a comprehensive summary of MOFs chemistry, isoreticular synthesis, and properties of isoreticular MOFs, synthesis advancements to tailor optical and electrical properties. The chapter mainly discusses the research advancement made towards investigating optoelectronic properties of IRMOFs. We also discuss the future prospective of MOFs for electronic devices with a proposed roadmap suggested by us. We believe that the MOFs-device roadmap should be one meaningful way to reach MOFs milestones for optoelectronic devices, particularly providing the potential roadmap to MOF-based field-effect transistors, photovoltaics, thermoelectric devices, and solid-state electrolytes and lithium ion battery components. It may enable MOFs to be performed in their best, as well as allowing the necessary integration with other materials to fabricate fully functional devices in the next few decades.

Keywords: MOF, coordination polymers, optoelectronics, isoreticular MOF, semiconducting MOFs

1. Introduction

Over last few decades, crystalline microporous materials, from zeolites, to coordination polymers and its subclass, metal organic frameworks (MOFs) have gained enormous attention in the scientific community due to their structural versatility and tailorable properties like nanoscale porosity, high surface area, and functional density [1, 2]. Metal organic frameworks have evolved in last few years as a revolutionary material that are self-assembled nanostructure [3, 4] built from metal ions and organic ligands. The first MOF, MOF-5 or IRMOF-1 ($Zn_4O(BDC)_3$) reported by Omar M. Yaghi was used in gas adsorption applications accounting to its high surface area of 2900 m^2/g [5, 6]. To date, 80,000 MOFs [7] have been reported owing to its diverse structure, compositions, tunable porosity, specific surface area, [8] ease of functionalization, unsaturated metal sites [9] and biocompatibility [10] . As a result, MOFs were used in a wide range of applications such as gas storage and separation, drug delivery and storage, chemical separation, sensing, catalysis, and bio-imaging [3, 7, 11–13]. In terms of structural orientation, the coordination bonding between a metal ion and organic ligand results in the formation of extended networks of one, two, and three-dimensional framework with potential voids [6, 14]. The coordination bonding facilitated through a suitable molecular approach, involving reticular synthesis, provides the flexibility to alter the pore size and transform its structure, targeting specific applications. Thus, utilizing the advantage of various combinations of metal-ligands and interaction of metal-ligands, MOFs are ideal candidates in the field of material science, offering an attractive property of structural tunability, providing a pathway to introduce and tailor intrinsic characteristics, such as optical, electrical, and magnetic properties.

There has been a growing interest exploring MOF as emerging semiconducting materials to meet the current demand in the electronic devices [15]. In particular, the electronic characteristics such as electrical, optical, and magnetic properties of MOFs have become an interesting topic of research attributing to their applications in microelectronic and optical devices. The implementation of MOFs in the electronic industry was first reported by Allendorf and co-workers [16]. MOF-5 with Zn_4O metal nodes and orthogonally interconnected six units of terephthalate is the most-studied MOF as a semiconductor. In 2007, Garcia and co-workers reported on the semiconducting behavior of MOF-5 synthesized at room temperature, with a bandgap of 3.4 eV [17]. Since then, intense research has been carried out to develop MOFs with semiconducting properties, opening new research domains for the scientific community in nanoscience.

The presence of narrow band gap structure either direct or indirect and charge mobility contribute to the semiconducting behavior of MOFs. To design MOFs with semiconducting behavior, significant amount of research is ongoing to identify the general structural requirements for enhancing the orbital overlapping between the building components. The main advantage of MOFs is the ability to tune the crystalline structure and functionality through phenomenal conceptual approaches such as rational designing and synthetic flexibility. In reticular chemistry, which is also known as rational designing, the coordination bonding between metal node and organic ligand provides an understanding of atomic positions precisely contributing to determine the fundamental structure–property relationships. Thus, the crystalline structure of MOFs consists of self-assembled ordered nanostructure with defined organized spatial space that is constructed via coordination chemistry between the building components.

Moreover, the sub-angstrom knowledge of atomic positions helps to eliminate any disorder in the structure that contributes to poor mobility in the structure. Considering synthetic flexibility, the electronic properties of MOFs could be tailored, resulting in potential applications such as a photovoltaic device tuned for solar cells, electroluminescent devices, field effect transistors, spintronic devices, and sensors. These developments have led many researchers to explore electrical, magnetic, and optical properties of MOFs [15, 18, 19]. However, the electrical properties of MOFs and integration of them in micro-electronic devices is still at an early stage and remain under research when compared to other types of existing conducting materials [4, 15] due to their insulating character. Although MOFs possess the properties of both organic and inorganic counterparts, they behave as electrical insulators or poor electrical conductors due to the poor overlapping between the π-orbitals of organic ligands and d-orbitals of the metal ion [20]. Yet, MOFs serving as an interface between (inorganic) hard and (organic) soft materials provide an opportunity for adapting various structure–property relationships that is related to wide range of parameters such as choice of metal ion, organic linker, and molecular designing approach. In general, the structure–property relationship in MOFs is a consequence of cooperative mechanism, i.e. the interaction between the metal and ligand, which could be readily identified by taking advantage of the knowledge of their detailed atomic structure, enabling fine tuning of their functionalities [7, 11]. According to the literature, Bastian Hoppe and his co-workers reported Cu-2, 3, 6, 7, 10, 11-hexahydroxytriphenylene (Cu_3hhtp_2-MOF), a copper-based graphene-like framework with inherent electrical conductivity about 0.045 S cm^{-1} [21]. MOFs with electrical conductivity higher than 0.1 S cm^{-1} was achieved by Talin and co-workers [22]. Thus, the designing of MOFs with conducting or semiconducting properties is necessary to enhance the sensitivity of electrical or demonstrate a sensing concept; but rarely have MOFs been an integral part of an actual device [23].

The purpose of this chapter is to provide comprehensive discussion on optoelectronic MOFs developed up to date and identify focus points to bring MOFs with optoelectronic properties for the realization of integrating MOFs into actual devices for electronic device applications. We first provide a MOFs chemistry and isoreticular synthesis advancements to make isoreticular MOFs (IRMOFs) with tailored optical and electronic properties. Then we summarize the current state of MOF research relevant to optoelectronics, particularly discussing the synthesis, electronic structure, and photophysical properties of three selected IRMOFs (IRMOF-1, 8, and 10). Finally, we propose a MOFs-device roadmap, focusing on MOF-based field-effect transistors, photovoltaics, thermoelectric devices, and solid-state electrolytes and lithium ion battery components.

2. Chemistry of MOFs

2.1 Dimensional classification and evolution of MOFs

Coordination polymers are organic–inorganic hybrid materials where organic moieties are bonded to metal ion or metal clusters via coordination bonds. The energy of such bonding is usually between 50 and 200 KJ mol^{-1}. Apart from strong coordination bonding, weaker interaction such as hydrogen bonds, van der Waal forces and π-π interactions also influence the formation of coordination polymers. Depending on the geometry, coordination polymers are classified into three sub-classes: (1) One-dimensional (1-D) coordination polymers, (2) Two-dimensional (2-D) coordination polymers, and (3) Three-dimensional coordination polymers (**Figure 1**).

The coordination polymer assembled from organic ligand and metal ion into three dimensional hierarchical crystalline structures is often regarded as metal organic framework. Since then, the term coordination polymer and metal organic

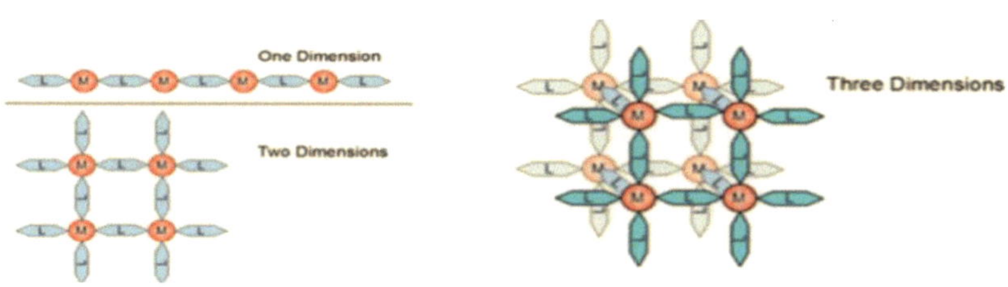

Figure 1.
Dimensional structures of coordination polymers.

framework have been used interchangeably. The term MOFs was first introduced by Omar Yaghi in 1995 [4, 9]. The framework of MOFs is either porous or non-porous. However, the porosity of MOFs was reported to be reversible due to various environmental factors (temperature, pressure, light intensity) contributing to the weak intermolecular interactions between building components. Thus, efforts have been made to modulate the strong structural rigidity that could incorporate permanent porosity. Based on this, in 1998 Kitagawa classified MOFs into three categories; 1st, 2nd, and 3rd generation coordinated network. Among three generations of coordinated networks, 3rd generation coordinated networks were defined to have permanent porosity with structural flexibility [10]. This led to numerous applications and implementation of coordinated networks in the gas storage community. The intermolecular interaction between organic ligand and metal ions, choice of building units, crystallization, environment, and guest molecules determine the crystal structural rigidity and dimensionality of MOF's coordination network. This major advance in the field of coordination polymer depicted that coordinated networks of MOFs could be modified and developed in a highly periodic manner, with a defined understanding of the crystalline structure, porosity, and chemical functionality. Thus, the ability to design and control the arrangement of metal ions with extended organic spaces in three-dimensional fashion led to the origin of the term reticular chemistry which was first introduced by Yaghi and coworkers [4].

With the variability of organic and inorganic components and their interaction, the freedom of modulating the structure of MOFs into highly ordered hierarchical structures with tunable pore volume and adjustable surface area has become feasible that made MOFs stand out compared to the other porous materials. Taking advantage of one of these hallmarks of MOFs i.e. designing of topologically diverse structures with desirable properties has been explored extensively attracting wide range of applications in gas storage, separation, catalysis, sensing and drug delivery [5]. Since 1990s, this area of chemistry has experienced tremendous growth in the field of material science and modern chemistry [4]. The flexibility with geometry, size, and functionality led to the "design" of a large number of MOFs. The organic units are generally ditopic or polytopic organic carboxylates, linked to metal-containing units, such as transition metals (e.g., Cu, Zn, Fe, Co, and Ni), alkaline earth elements (e.g. Sr., Ba), p-block elements (e.g. In, Ga), and actinides (e.g. U, Th) [6]. A major advance in the chemistry of MOFs came in 1999 with the invention of two structures i.e. MOF-5 (IRMOF-1) and HKUST-1 [11] reported by Omar et al. and Chui et al., respectively. Subsequently, in the coming years around 2002, the flexible and non-flexible structures of MIL-88/53 [12] was reported by Ferey et al.

2.2 Reticular chemistry and isoreticular MOFs

The demand for the synthesis of new materials to perform highly specific and cooperative functions has been increasing rapidly in parallel with advanced technology [13, 14]. Recently, the field of metal organic framework has evolved significantly due to its practical and conceptual approach to design and develop the target material. Intrinsically, the reticular chemistry is described as the process of assembly of molecular building blocks held together by strong bonding that pattern into periodic arrays of the ordered net like structures [13–16]. Some of the advantages of this approach are: (1) Molecular approach, which provides the ability to design and control the structure of frameworks [17]; (2) Bonding in which the strong bonding between the building blocks could impart superior functionalities like thermal and chemical stability into the framework; and (3) Engineered crystallinity, which is based on the type of the interactions (intermolecular or intramolecular) design and synthesis with controlled and desired properties.

After the introduction of the parent MOF, MOF-5, taking advantage of reticular chemistry that includes reticulating metal ions and organic carboxylate, the group of Omar M. Yaghi synthesized a new class of materials called IRMOFs. Thus, the theory of isoreticular chemistry was established in the year 2002 with the development of IRMOFs. These class of materials were developed to improve the surface area and pore volume by incorporation of different topological linkers. In IRMOF, IR stands for isoreticular, which means it is a series of MOFs with the same topology, but different pore size [14, 20, 22, 23]. A series of different IRMOFs share similar *pcu* topology of IRMOF-n (n = 1–16). As shown in the **Figure 2**, the pore volume and porosity vary with the variation in the organic linker. Applying the concept of isoreticular chemistry, various kinds of MOFs were developed.

2.3 Synthetic advancements of MOFs

The conceptual approach of designing and assembling a metal–organic framework follows reticular synthesis and is based upon identification of how building blocks come together to form a net, or reticulate. The synthesis of MOFs is often regarded as "design" which implies to construct, built, execute, or create according to the target material. The synthesis approach for a new MOF should follow several factors asides from the geometric principles that are considered during its design. Among such factors, by far the most important is the maintenance of the integrity of the building blocks. A great deal of research effort has been demonstrated on the synthesis of a novel organic link and synthesis conditions that are mild enough to maintain the functionality and conformation of organic ligand, yet reactive enough to establish the metal–organic bonds. In situ generation of a desired secondary subunit (SBU) is required carful design of synthetic conditions and must be compatible with the mobilization and preservation of the linking units [24]. Typically, this is achieved by precipitation of the product from a solution of the precursors where solubility is a necessary attribute of the building blocks but is quite often circumvented by using solvothermal techniques [24].

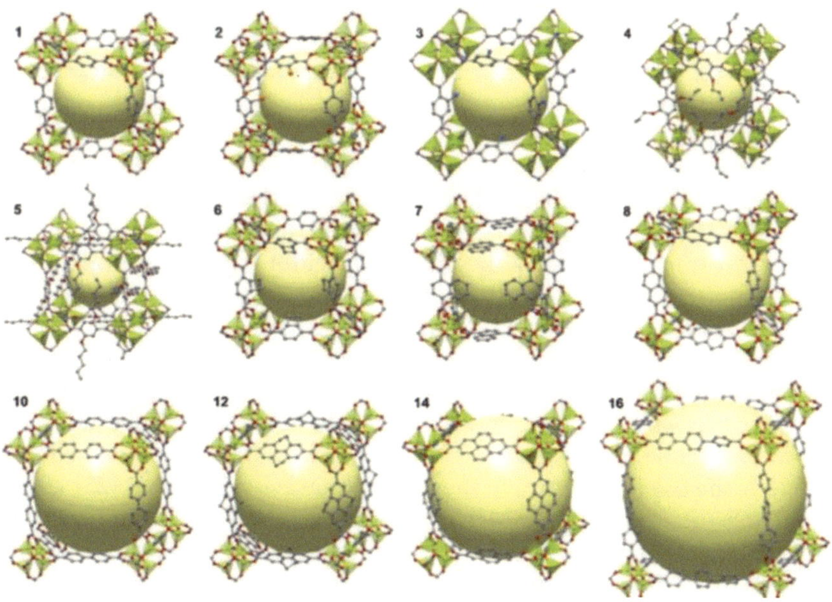

Figure 2.
Crystal structures of IRMOFs-n series [n = 1–16]. The non-interpenetrated structures from (n = 1,2,3,4,5,6,7,8,10,12,14,16). The yellow sphere represents the pore volume. Zn atoms are in green, O in red, C in gray, Br atoms in Orange, and amino groups in blue [17].

Traditional goal of MOF synthesis is to obtain high quality single crystal for deducing the structure and understand the crystal packing, geometry, and pore volume with respect to the organic ligand's length. Thus, prior to begin elucidating the concept of reticular synthesis, most early studies were exploratory and early stage synthesis has mainly involved simple, highly soluble precursors, and labile metal ions of the late transition series. The polymerization that leads to make 3D-network of MOFs needs an assembly process where an insoluble entity is quickly formed that precludes recrystallization. Fortunately, it differs in the degree of reversibility of the bond formation event, allowing detachment of incoherently matched monomers followed by reattachment with continued defect free crystal growth. The framework assembly occurs as a single synthetic step, where all of the desired attributes of the target material constructs by the building blocks. This often requires a combinatorial approach, which involves subtle changes in concentration, solvent polarity, pH, or temperature. Any subtle changes in these parameters leads to poorer quality crystals, reduced yields, or the formation of entirely new phases [24].

Augmenting simple crystal growth processes used to grow simple inorganic salts, early efforts of producing highly crystalline MOFs involved the slow introduction of the building blocks to reduce the rate of crystallite nucleation. Methods included slow evaporation of a solution of the precursors, layering of solutions, or slow diffusion of one component solution into another through a membrane or an immobilizing gel [24]. During the nucleation stage, the ligand deprotonation prior to the coordination onto metal ion is catalyzed introducing a volatile amine gradually via vapor diffusion. Just as for many of the polar solvents used, suitable choice of base is necessary to avoid competitive coordination with the organic links for the available metal sites. While in some cases, blocking of metal coordination sites is necessary for the formation of a particular SBU. However, this approach has generally been regarded as leading to low-dimensional structures that are less likely to define an open framework.

With the need for more robust frameworks, having larger pore volumes and higher surface area, introducing bulker, longer length organic linkers are necessary, but greater difficulties in crystal growth were encountered. Thus, later, solvothermal techniques were found to be a convenient solution to overcome this challenge and have largely benefit over often time-consuming methods involving slow coupling of the coordinating species. The typical solvothermal method combines the precursors as dilute solutions in polar solvents such as water, alcohols, acetone or acetonitrile and heated in sealed vessels such as Teflon-lined stainless-steel bombs or glass tubes, generating autogenous pressure. The crystal growth process is enhanced by using mixed solvent systems where the solution polarity and the kinetics of solvent-ligand exchange can tune to achieve rapid crystal growth. It has found that, exposing the growing framework to a variety of space-filling solvent molecules may also be an effective way to stabilize its defect-free construction as they efficiently pack within the defined channels [24]. For deprotonation of the linking molecule alkyl formamides and pyrrolidinones have been particularly useful, as they are also excellent solubilizing agents.

In recent years, modifying the solvothermal method, there are several rapid synthesis methods were proposed by researchers to develop MOF crystals within a short duration of time. Some of the external parameters implemented to develop MOFs include the use of Microwave energy (Microwave synthesis), [25] Ultrasonic waves (Sonochemical synthesis), Mechanical energy (Mechano-chemical synthesis) and electrical energy (Electrochemical synthesis). The synthetic strategies developed up to date to make different type of MOFs are summarized in the **Table 1** along with reaction conditions [26]. Additionally, a surfactant driven-templating method, [22] a CO_2-expanded liquid route, [27] a post-synthetic method, [28] and an ionic liquid-based method [29] are developed to create hierarchical mesoporous microstructures and thin films of MOFs [25, 27–29].

Synthesis method	Reaction time	Temperature (K)
Slow evaporation	7 days to 7 months	298
Sonochemical method	30–180 mins	272–313
Solvothermal method	48–96 hours	353–453
Mechano-chemical method	30 min to 2 hours	298
Electrochemical method	10–30 mins	273–303
Microwave Synthesis	4 mins to 4 hours	303–373

Table 1.
Synthesis methods developed up to date to make MOFs.

3. $Zn_4O(L)_3$-based isoreticular MOFs with cubic topology for optoelectronics

3.1 Road map to electrically conductive MOFs

In the area of MOFs, the main desire is to design MOFs with optoelectronic properties and to optimize the charge transport mechanism suitable for developing electronic devices. Although numerous applications of MOFs with different types of synthesis methods are being investigated, a versatile and scalable synthesis approach for the preparation of MOFs with semiconducting properties for optoelectronic devices are still in the early stage and a little research work so far done towards tailoring MOFs structure–property relationship to use as active materials in optoelectronic devices, such as solar cells, field-effect transistors, and photoluminescence devices.

To introduce MOFs as semiconducting materials, tuning of band gap such as lowering the bandgap or increasing the charge mobility is required. This tunability is again dependent upon the type of interaction i.e. Intermolecular interaction: metal ion and the organic ligand or Intramolecular interaction - π stacking [18]. The two key factors responsible for poor electrical conductivity in MOFs are: (1) the insulating character of organic ligand and (2) due to poor overlapping between the π-orbitals of organic ligand and d-orbitals of metal ions [16]. The common strategies for constructing MOFs with electrical conductivity involves three possible charge pathways.

Pathway 1: A long range of charge transport in this pathway is facilitated through bonds. This mechanism is promoted by interaction between ligand π and metal d orbital [16]. This mechanism is based on the tunneling of electron between the donor and acceptor portions of the framework. Typically, the electrical conductivity in the range 10^{-7} to 10^{-10} S cm^{-1} is considered as insulator. This is caused due to poor overlapping between the metal ion and organic linker as the electronegative nature of oxygen atom in the carboxylate group of the linker is so high that it requires high voltage for tunneling of the electrons [30]. Various MOFs that exhibit conductivity through this mechanism have been reported, of which $[[Cu_2(6\text{-Hmna})\ (6\text{-mn})]\cdot NH_4]_n$, a copper-sulfur based MOF constructed from 1,6-Hmna = 6-mercaptonicotinic acid, 6-mn = 6-mercaptonicotinate shows highest electrical conductivity of 10.96 S/cm (**Table 2**).

Pathway 2: In this pathway, the charge transport is facilitated through space via π stacked aromatic ligands which was proposed as an alternative to through bond strategy. This mechanism typically promotes electron hopping mechanism by employing electroactive molecules [16, 30]. TTF-TCNQ i.e. tetrathiafulvalene-tetracyano quinomethane is one of the MOFs that demonstrate metallic conductivity

Materials	Mechanism	Conductivity (Scm^{-1})	Charge carrier	Mobility (cm^{2}V^{-1} S^{-1})	Ref.
Metals Cu Au Fe	Tunneling	6.5 x105 4.1x105 1.0 x 105	e	46	[31]
Organic polymers Polyacetylene Polythiophenes Rubrene	Charge transfer (electron hopping)	10–9 1975	h	1–10 4	[31]
TTF-TCNQ Ni_3 (HITP)$_2$ Zn_2 (TTFTB)	Through-space	700 40 4.0x10–6	h or e	48.6 0.2	[16, 31, 32]
Cu_3(BTC)$_2$-TCNQ NU-901-C60	Guest molecule	0.07 1x10–3	h		[33, 34]
Fe_2(DSBDC) {[Cu_2(6-Hmna) (6-mn)] ·NH_4]$_n$ Cu [Ni(pdt)2]	Through-bond	1x10–6 10.96 1x10–4			[18, 35–37]

Table 2.
Significant progress in the last few years made towards developing electrically conductive MOFs and their conductive properties compared with conventional metals.

(shown in the **Table 2**) through-space (π-π stacking) mechanism [38]. Recently, TTF-based ligand consisting of benzoate spacers is used to develop Zn based MOF reported by Dincă et al. These MOFs shows columnar stacks of TTF (3.8 Å) with the charge mobility of a magnitude that resembles some best conductive organic polymers [35, 36].

Pathway 3: The other alternative strategy to increase the conductivity of MOFs is via incorporating an appropriate guest molecule within the MOF. These molecules can activate long range delocalization either through bonds or through space or that can inject mobile charge carriers by oxidizing or reducing the organic ligand and metal ions [16, 30] NU-901, a MOF consisting of Zr_6 (μ_3-O)$_4$ (μ_3-OH)$_4$ (H_2O)$_4$ (OH)$_4$ nodes and tetratopic 1,3,6,8-tetrakis (p- benzoate) pyrene (TBAPy4-) linkers. These materials were chosen for the encapsulation of C_{60}. After installation of C_{60}, the NU-901-C_{60} shows electrical conductivity higher than that of NU-901 (shown in the **Table 2**). As per reports, the donor-acceptor interactions between TBAPy$_4$-/C_{60} contribute to the electrical conductivity of the framework [32, 38].

3.2 Synthesis and optoelectronic properties of IRMOF-1

IRMOF-1, which is commonly known as MOF-5, invented by Yaghi and co-workers [39] in 1996, has become one of mostly studied MOF with promising application in high capacity hydrogen storage [40, 41]. MOF-5, consists of Zn_4O units connected by linear 1,4-benzenedicarboxylate units to form a cubic network, having the primitive cubic unit cell. Syntheses demonstrated for MOF-5 in which the starting materials are mixed in solution at ambient temperature. Subsequent addition of triethylamine promotes the deprotonation of the organic linker to precipitate MOF-5. Depending on the addition of base either slowly by diffusion as described in the original synthesis method [39] or rapidly as an aliquot [42] the product can be either single crystal mixtures, which must be mechanically separated, or microcrystalline powders. The ambient temperature synthesis method described above following the fast addition of base, is easy to scale up. However, metal precursor, zinc nitrate poses potential safety concerns, especially for large-scale production. Furthermore, reports of such synthetic conditions have been largely limited to MOF-5 and IRMOF-8 [42–44].

Later, a rapid, simple, room-temperature high yielding synthesis method was introduced by Yaghi and co-workers that can apply to make a wide range of new MOFs, including IRMOF-0, which uses acetylenedicarboxylate as the linker [45]. This synthesis method follows a room temperature synthesis, wherein separate N,N-dimethylformamide (DMF) solutions of terephthalic acid (BDC) with triethylamine and zinc acetate dihydrate are prepared, then the zinc salt solution is added to the organic solution with rapid stirring at ambient temperature. Upon immediately of the formation of a white powder followed by 2.5 hours of reaction time, pure MOF-5 is collected and confirmed by powder XRD. The same synthesis without a base (triethylamine) has also yielded pure MOF-5, confirming that addition of a base is unnecessary when zinc acetate is used as a source of Zn (II) in the MOF-5 synthesis [45].

This synthesis method has later modified by Rathnayake et.al to make IRMOFs (IRMOF-1, 8, and 10) by cutting down the reaction time from 2.5 hours to 7–9 minutes [23]. As depicted in **Figure 3**, our group is able to make a wide range of hierarchical microstructures of highly crystalline MOFs, including IRMOF-1. Microstructures of IRMOF-1 prepared from the modified solvothermal method (**Figure 3**), are visualized using scanning electron microscope and are depicted in **Figure 4(a)**. Crystal structure of IRMOF-1, retrieved by matching its simulated XRD with experimental powder XRD is depicted in **Figure 4(b)**, and follows cubic lattice cell, which belongs to Fm3m cubic space groups. The electron density potential distribution modeled from VESTA (**Figure 4(c)**) evidences that the electron potential is localized on Zn_4O clusters and there is no electron delocalization with the organic linkers, confirming no orbital overlap for energy transfer through metal–ligand charge transfer processes.

As a first member of isoreticular series, IRMOF-1 has explored for luminescence due to ZnO quantum dots behavior, which has been believed, contributing to luminescence. The ZnO QD absorption and emission spectra from electronic transitions have been investigated, suggesting that that the luminescent behavior of IRMOF-1 arises from a $O_2 - Zn^+ \rightarrow O - Zn^+$ charge-transfer transition within each tetrahedral Zn_4O metal cluster, which has been described as a ZnO-like QD [46]. The photoluminescence emissions of IRMOF-1 with intensity peak maximum at 525 nm, was ascribed to energy harvesting and LMCT from 1,4-benzenedicarboxylate (BDC) linked to the Zn_4O cluster. The nature of the luminescence transitions in IRMOF-1 nanoparticles has been investigated by Tachikawa et al. where the

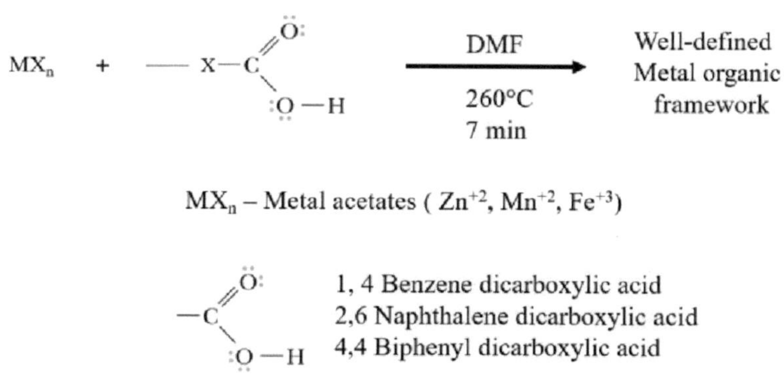

Figure 3.
Reaction scheme for synthesis of Isoreticular MOFs using modified solvothermal method followed by solvent driven self-assembly.

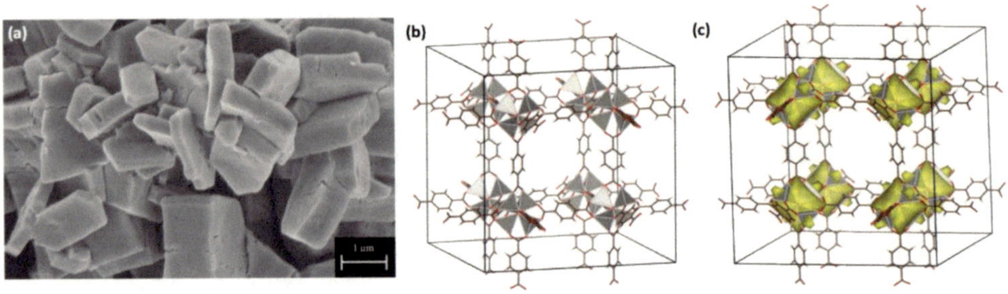

Figure 4.
(a) A SEM image of IRMOF-1 microstructures, (b) crystal structure of IRMOF-1 retrieved from crystallographic open database, and (c) electron density potential distribution of IRMOF-1 modeled from VESTA.

transition responsible for the green emission of IRMOF-1 is similar to that of ZnO [47]. Therefore, the emission observed in IRMOF-1 has been speculated to originate from the ZnO QD not from the ligand. However, Further investigations demonstrated that ZnO impurities in the material gave rise to the emission assigned to the quantum dot like luminescence and that pure MOF-5 displays a luminescence behavior that is more closely relevant to that of the ligand. [9] However, the exact nature of the luminescence of MOF-5 is still under dispute with ligand– ligand charge transfer, [10] ligand-centered, [9] and ligand–metal charge transfer [11] mechanisms as primary suggestions.

In an on-going study, our group has been investigating optoelectronic behavior of IRMOF-1. As depicted in **Figure 5**, UV–vis absorption spectrum shows absorption vibrionic features similar to the linker with two absorptions peaks at 208 nm and 240 nm along with a shoulder peak at 285 nm. The emission spectrum collected by exciting at 240 nm exhibits linker-based emission with three well-resolved vibrionic transitions at 328 nm, 364 nm, and 377 nm. We observed a small high energy shoulder peak at 464 nm, which corresponds to an excitonic transition of Zn_4O nodes. However, we have no observed a longer wavelength emission peak at 525 nm, which has claimed in prior studies to energy harvesting and LMCT from 1,4-benzenedicarboxylate (BDC) linked to the Zn_4O cluster. Therefore, our findings support that IRMOF-1's luminescence comes from linker emission rather than the charge transfer processes. This further excludes the emission originating from the ZnO quantum dots like clusters of Zn_4O. The optical band gap calculated from the UV–visible spectrum on-set is found to be 3.97 eV. There are no experimental band gaps reported for IRMOF-1 up to date.

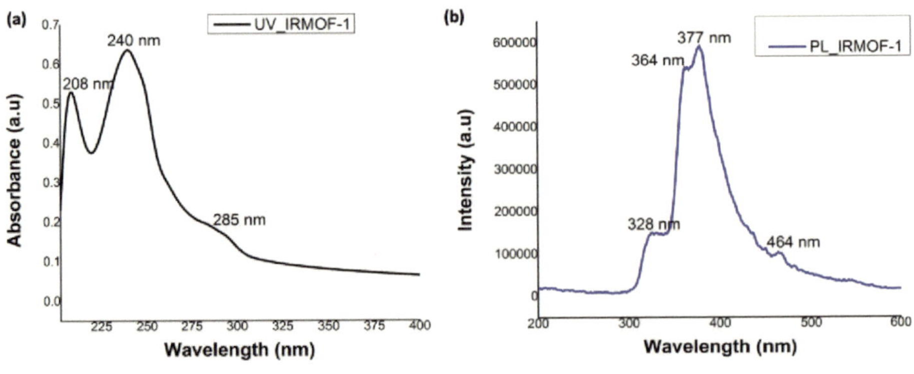

Figure 5.
Photophysical properties of IRMOF-1 – (a) UV–visible spectrum and (b) photoluminescence spectrum in solution (ethanol).

3.3 Synthesis and optoelectronic properties of IRMOF-8

Significant research efforts have demonstrated successful synthesis of a variety of isoreticular MOFs (IRMOFs) with the formula of $Zn_4O(L)_3$ (where L is a rigid linear dicarboxylates) using traditional solvothermal method, which uses zinc nitrate as metal precursor and the respective organic ligands in an amide-based solvent system. These IRMOFs follow the same cubic topology as the prototypical MOF-5, a framework with octahedral $Zn_4O(CO_2)_6$ clusters, which are linked along orthogonal axes by phenylene rings [3, 26, 48, 49]. This family of IRMOFs-n (n = 1–16) gained significant attention in gas storage community due to its high pore volume and surface area. Among the IRMOFs series, IRMOF-1 and 8 have been extensively studied for gas adsorption and photoluminescence properties [39, 50, 51] but have not explored their optoelectronic properties until recently.

IRMOF-8 is constructed from the linkage of basic zinc acetate clusters and naphthalene-2,6-dicarboxylic acid units (NDC). Originally reported IRMOF-8 with non-interpenetrated cubic crystal lattice has only been extensively studied for gas sorption and storage applications [50, 51]. Later, a number of interpenetrated phases of $Zn_4O(ndc)_3$-based systems have been discovered [52–54]. Although the synthesis of interpenetrated IRMOF-8 (INT-IRMOF-8) are similar to that of IRMOF-8, the possibility that typical solvothermally synthesized IRMOF-8 contains at least a significant amount of an interpenetrated phase. There are modified synthesis methods have been introduced to make fully non-interpenetrated IRMOF-8 [55] and INT-IRMOF-8 [23, 55]. The crystal structures of non-interpenetrated IRMOF-8 and INT-IRMOF-8 along with their space filling structures, acquired from the Crystallographic Open Database (COD) and generated using VESTA are depicted in **Figure 6**.

Recently, our group has introduced a modified solvothermal synthesis method, which involves a solvent polarity driven self-assembly process to make hierarchical microstructures of INT- IRMOF-8, exhibiting promising optoelectronic properties for the first time [23]. Instead using zinc nitrate as the metal precursor, the synthesis we developed utilizes zinc(II)acetate as the metal precursor. Hierarchical microstructures of INT-IRMOF-8 nanocrystals can be prepared in high yield in the presence of minimum volume of dimethyl formamide by mixing zinc(II) precursor with naphthalene-2,6-dicarboxylic acid at room temperature followed by subjecting to solvothermal annealing at 260°C for 7 minutes [23]. Microstructures visualized under TEM (**Figure 7(b)**) reveal that they are hierarchical layers of self-assembled nanocrystals with randomly arranged voids among the nanocrystals. The wide-angel X-ray scattering (WAXS) pattern along with the selective area electron diffraction (SAED) pattern have shown that the microstructures are made from self-assembled nanocrystals of INT-IRMOF-8, which exhibits lamella packing pattern (**Figure 7(c)** and **(d)**), benefiting for optoelectronic behavior.

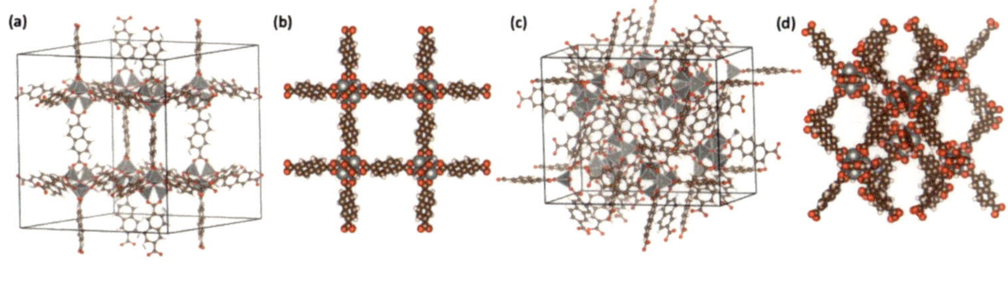

Figure 6.
Crystal structures of: (a) non-interpenetrated IRMOF-8 and (b) its space filling view, (c) INT-IRMOF-8 and (d) its space filling view.

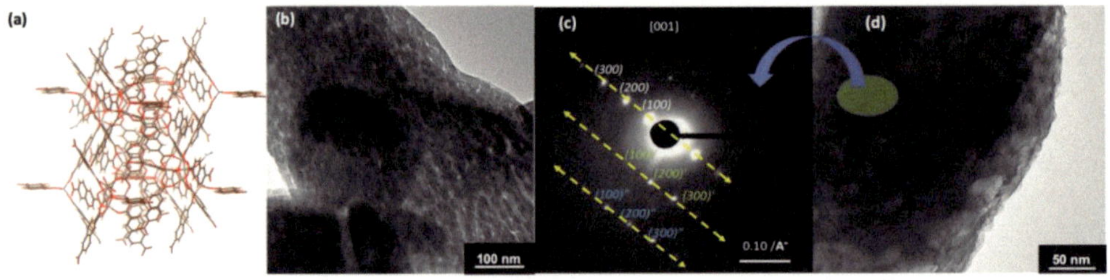

Figure 7.
*(a) Interpenetrated view of INT-IRMOF-8's crystal structure; (b) a TEM image of a microstructure; (c) the SAED pattern of a microstructure taken from the TEM under dark field diffraction mode along with (d) a TEM image of the respective microstructure. [**Figure** 7 is re-created from the original data].*

The photophysical properties of INT-IRMOF-8 exhibit mainly linker based optical properties. The presence of high intensity absorption peak set from 220 nm to 360 nm, which corresponds to vibronic π-π^* absorption transitions of naphthalene core, evidencing the linker-based absorption, resulting from the lack of favorable spatial and energetic overlap of the metal and the ligand orbitals [21, 49]. Typically, MOFs' photoluminescence behavior arises as a result of different types of charge transfer processes, which include metal-to-ligand charge transfer (MLCT), ligand-to-metal charge transfer (LMCT), ligand–ligand charge transfer (LLCT), ligand-centered luminescence, and metal-to-metal charge transfer (MMCT) processes [56]. However, this metal-centered luminescence depends on the metal type, ligand type, and their spatial orientations. The emission spectrum of INT-IRMOF-8 microstructures exhibits three emission bands, in which vibrionic transitions corresponds to only linker-based emission with no indication of additional emissions due to any charge transfer processes. The optical band gap reported for INT-IRMOF-8 is 2.82 eV [23] and the theoretical band gap reported in the past for non-interpenetrated IRMOF-8 was ranged from 2.83 eV to 3.27 eV [57]. There are no experimental band gaps reported for IRMOF-8 up to date.

The charge transfer ability of IRMOF-8 for the first time is evaluated by our group. The average electrical conductivity of INT-IRMOF-8 microstructures was ranged from 3.98×10^{-2} to 2.16×10^{-2} S cm^{-1}, which is higher than the electrical conductivities reported for most MOFs ($<10^{-10}$ S/cm). The interpenetrated structure, high crystallinity, and narrow band gap contribute to the to the electrical conductivity of hierarchical structures of INT-IRMOF-8 nanocrystals.

3.4 Synthesis and optoelectronic properties of IRMOF-10

Among the series of IRMOFs (IRMOF 1–16) introduced by Yaghi and coworkers [58–60], several IRMOFs have shown effective selective preconcentration properties, including IRMOF-10 [61–64].

Compared to IRMOF-1, physicochemical, optical, and electronic properties of IRMOF-10 with its 4,4′-biphenyldicarboxylate linkers has received much less attention. IRMOF-10 was first synthesized by Yaghi and coworkers [50, 58–60], Owing to its higher surface area and larger pore sizes, use of IRMOF-10 for gas absorption and separation and hydrogen storage have been widely investigated, but scarce attention has been paid to other properties of IRMOF-10, such as structural stability, optical and electrical properties, electronic structure, and chemical bonding. The first publication about biphenyl MOFs already anticipated the major challenges related to Zn-biphenyl MOFs: the growth of single-crystals and interpenetration. A structure from single-crystal XRD of non-functionalized IRMOF-10 is not yet available. A single-crystal X-ray structure analysis of a non-interpenetrated

IRMOF-10 derivative was not reported until the breakthrough of the group of Telfer, which showed how interpenetration can be effectively suppressed by using thermolabile protecting groups in the synthesis of amino-MOFs [65]. Following the modified solvothermal synthesis method introduced by Rathnayake et al., microstructures of non-interpenetrated IRMOF-10 was successfully synthesized, and crystal structure was confirmed by matching the powder XRD traces with the simulated XRD pattern. The microstructures morphology is depicted in **Figure 8(a)** and crystal structure retrieved from the Crystallographic Open Database is depicted in **Figure 8(b)**. IRMOF-10's single crystal structure reveals three-dimensional coordination framework, formed by periodic arrangement of Zn(II) atoms, which is tetrahedrally coordinated by four oxygen atoms from four biphenyl linker units, following the unit formula of $Zn_4O(L)_3$ with cubic topology as IRMOF-1.

The electron potential density localization surrounding the metal oxide nodes and organic linker units in IRMOF-10's unit cell reveals the electron density distribution with respect to the biphenyl conjugation length. As shown in **Figure 8(c)**, the electron potential is delocalized within metal oxide nodes and bi-phenyl units, and partial distribution of charges has increased around bi-phenyl units compared to naphthalene units of IRMOF-8. Thus, the findings suggest that linker length has more pronounced effect on the individual material's electronic band structure and density of state, providing clear visualization on the localization of electronic potential within the crystal lattice. The delocalization of electron density potential through biphenyl linkers evidences its potential to be used as optoelectronic materials. Thus, exploring its electronic structure, band gap, optical, and electrical properties is a major interest to the materials science community. However, despite computational investigations on theoretical prediction of optoelectronic properties, [66] there are no experimental investigations on IRMOF-10's optoelectronic behavior has been conducted up to date.

The equilibrium solid-state structure, electronic structure, formation enthalpy, chemical bonding, and optical properties of IRMOF-10 have investigated with density functional calculations. Electronic density of states and band structures study have shown that the band gap for the IRMOF-10 is ranged from 2.9 eV to 3.0 eV, resulting in a nonmetallic character [66]. Until now, there are no experimental studies available to verify theoretical predictions on IRMOF-10's electronic structure. The calculated optical properties of IRMOF-10 provide useful information for future experimental exploration. The optical properties (dielectric function, refractive index, absorption coefficient, optical conductivity s(v), reflectivity, and electron energy-loss spectrum of the IRMOF-10 have computed in the past, [66] but there is no experimental investigation conducted up to date.

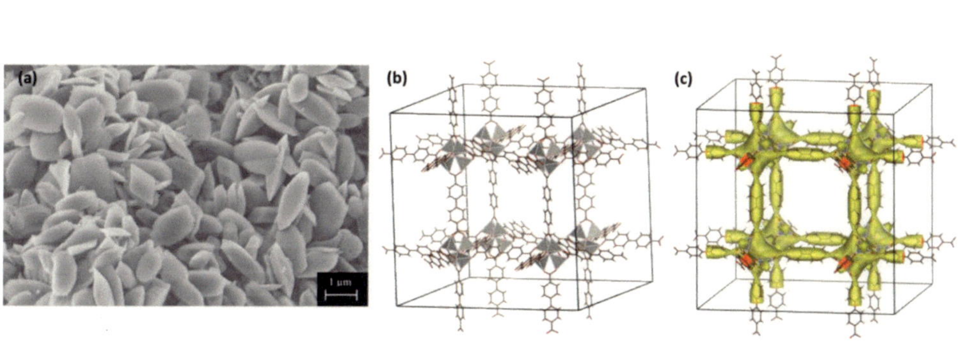

Figure 8.
(a) A SEM image of IRMOF-10 microstructures, (b) crystal structure of IRMOF-10 retrieved from crystallographic open database, and (c) electron density potential distribution of IRMOF-10 modeled from VESTA.

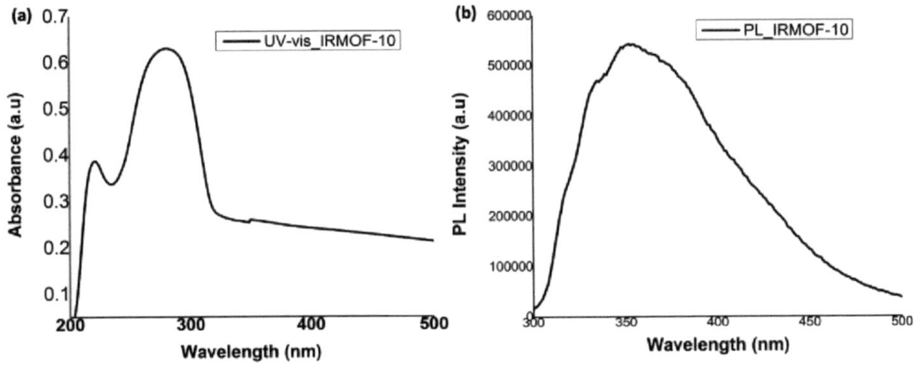

Figure 9.
Photophysical properties of IRMOF-10 – (a) UV–visible spectrum and (b) photoluminescence spectrum in solution (ethanol).

Recently, our group has studied optoelectronic behavior of non-interpenetrated IRMOF-10. As shown in **Figure 9**, we explored the photophysical properties of non-interpenetrated IRMOF-10 and calculated its optical band gap. IRMOF-10 exhibits linker-based absorption with absorption maximum at 282 nm along with a shoulder peak at 222 nm. IRMOF-10 shown blue luminescence with broader emission ranged from 310 nm to 450 nm along with the emission maximum at 353 nm. The optical band gap calculated from the UV–visible spectrum on-set is 3.80 eV, which is narrower than the optical band gap of IRMOF-1 and larger than the theoretical band gap predicted from computational analysis.

In summary, the conjugation length of the organic linker in IRMOFs contributes to the semiconducting properties rather than the periodic pattern or the distances between the Zn_4O clusters. The conjugation length of organic linkers of IRMOF-1, 8, and 10 described here differs from one aromatic unit (benzene) to one and half aromatic unit (naphthalene) to two aromatic units (biphenyl). The resonance effect arises due to the conjugation speaks directly to the photophysical behavior and optical band structure characteristics, reflecting a clear trend in narrowing the band gap with gradual increase in the conjugation length of the ligand. The dramatic change in the optical band gap upon changing the organic linker in the MOF structure has also been reported in the past [66]. Thus, these studies evidences that the semiconducting properties of MOFs strongly depends on the resonance effects from the organic linker [67].

4. Future prospective

With a growing demand for continuous miniaturization and functional scaling, the complementary metal-oxide semiconductor (CMOS) platform continues to drive advances in integrated circuits (IC), nanoelectronics and information processing technologies. While it is now possible to produce an amazing array of nanoscale materials and morphologies, the assembly and integration of these nanostructures into ordered arrays, with other materials, remain key challenges. Moore's Second Law projects a need for new, high throughput fabrication approaches that can produce useful and defect free nanostructures for future silicon-based CMOS related technologies. Recent advances in nanomaterial synthesis enable new families of emerging research materials (ERMs) that show potential for extending and augmenting existing CMOS technology, with respect to wafer level manufacturability, uniformity, reliability, performance and cost, and they warrant additional research focus and verification. The integration of More-than-Moore, application specific, materials and structures on a CMOS platform leverages the best of both technologies, though this added complexity also challenges the extensibility of conventional fabrication and patterning methods. Consequently, there remains a need for simple fabrication methods that can create two- and three-dimensional ordered functional nanostructures, which can adapt to a wide variety of materials, patterning, and application needs.

Highly crystalline microstructures of MOFs have been paving the path, addressing the current challenges in fabrication needs that create two- and three-dimensional ordered structures and which are adaptable to a wide variety of materials specific applications. These nanoscale building blocks, and their assemblies combine the flexibility, conductivity, transparency, and ease of processability of soft matter (organic) with electrical, thermal, and mechanical properties of hard matter (inorganic). They offer a new window for fine-tuning structural nodes with known geometries and coordination environments. With respect to the fabrication of ordered nanoscale structures, MOFs have several advantages. First, since they are themselves a highly ordered self-assembled nanostructure, as a result of their crystallinity, their pore dimensions are completely defined, making knowledge of atomic positions possible. Second, the nanoporosity of their structure results from geometric factors associated with the bonding between their inorganic and organic components, enabling rational template design [68]. Third, unlike the conventional template materials, MOFs possess a high degree of synthetic flexibility with potentially widely tunable electrical, optical, and mechanical properties. Surely, the development of simple, versatile low-cost methodologies for the design, production, and nanoscale manipulation of innovative functional organic–inorganic hybrid building blocks will provide a powerful new capability for designing, integrating, and patterning new nanoscale materials with tunable properties onto a CMOS platform. Recent milestones of MOFs in photovoltaic, optical and chemical sensing, and field effect transistors highlight the potential of these materials for future electronic devices, heterogeneous platforms, non-traditional patterning opportunities [16, 69–71].

Interest in using these materials in fields such as gas storage [72], separations [73], [sensing [21], and catalysis [74] is rapidly accelerating. The advantages of MOFs for above applications are promising due to the wide range of possibilities of the rational design inherited in these materials. Thus, superior properties and new understanding with respect to the interaction of small molecules with nanoporous materials are being achieved. Although most MOFs are found to be dielectrics, a few semiconducting frameworks are known [23, 37, 75, 76]. The theoretical predictions conduced up to date on variety of MOFs suggest that there are possible MOFs with semiconducting properties [77–79] . MOFs that are magnetic [80], ferroelectric [81, 82], proton-conducting [83–86], and luminescent [87, 88] are also known. Additionally, their porosity creates the potential to introduce non-native functionality to a given structure by infusing the accessible volume with a second molecule or material. Moreover, because the chemical environment within the pore can be modified, it is possible to tailor the interface between the MOF and a templated material to stabilize specific materials or nanostructures. Consequently, MOFs and the coordination polymers of crystalline nanoporous frameworks possess many of the properties of an ideal template.

Despite the endless possibilities for how MOFs could be used for device applications, when using MOFs for semiconductor microelectronic devices such as sensors, field-effect transistors, light harvesting and absorbing, light-emitting diodes, thermoelectric devices, energy storages and lithium ion batteries, and scintillators, it is necessary to understand how these materials function within the device and how they will interface with other functional and structural elements. Therefore, this section focuses on providing a future prospective for advances that must be made for their realization in electronic devices. A possible MOF-device roadmap, which identifies MOFs applications in electronic devices along with machine learning for new MOFs developments and MOFs database screening for novel properties is depicted in **Figure 10**. Our intention of providing this roadmap is to stimulate future endeavors of MOFs roadmap for electronic industry by translating current MOFs basic research agenda into applied research in the future. The roadmap that we identified here is created by combining the prospective previously provided by Allendorf et al., focusing five major fields pertinent to device fabrication [89]. These previously proposed areas include: (1) Fundamental Properties, (2) Thin film growth and processing, (3) MOFs hybrid and multilevel structures, (4) Device integration, and (5) Manufacturing issues. Our prospective for the proposed potential MOF-device roadmap particularly concentrated on member-specific applications in electronic industry, where functional density of MOFs can be utilized in subcategories of a wide variety of electronic devices. As the MOF-based optoelectronic field is fairly new and fall within the basic research stage, our roadmap is structured based on MOFs relative progress made so far and build upon the future road map comparing with the progress made in the field of organic electronics.

Exploring electronic properties, such as electronic structure, band gap, conductivity, electron, and hole mobilities, and dielectric constants of MOFs need to be one of the priority areas in the next decades and must be understood. Additionally, understanding lattice defects and their relationship to electronic properties must be explored combining theoretical and experimental approaches as they likely will limit the ultimate performance of a device. The field-effect transistor (FETs), which is the basic device building block for modern electronics, dictates the materials properties relevant to electronic applications. The FET performance is determined by the carrier mobility, source and drain contact resistance, and the capacitance of the gate electrode. Si is the preeminent materials for FET fabrication because of its bandgap of 1.1 eV, high carrier mobility, availability of multiple n- and p-type dopants, environmental stability, stable oxide, and high terrestrial abundance. However, Si based device fabrication requires enormous capital investment and Si is not compatible with a variety of low cost, flexible, transparent, and low melting

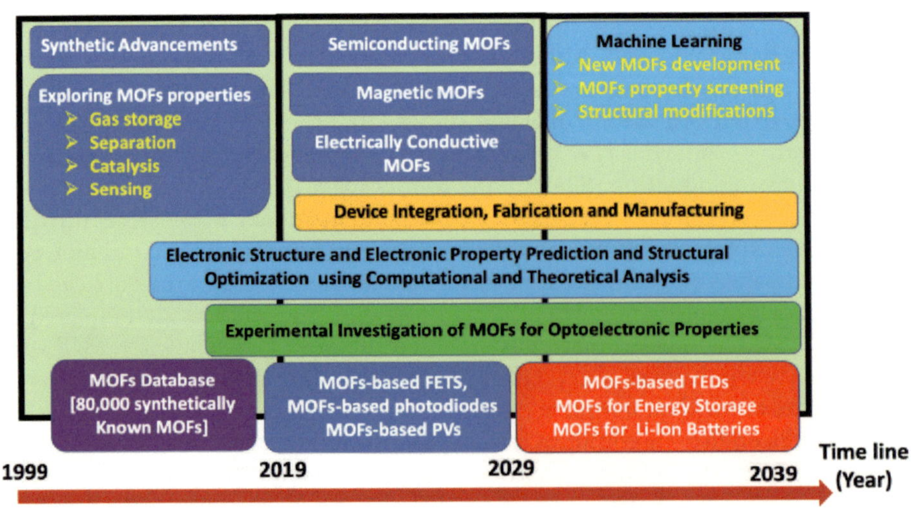

Figure 10.
A possible MOF-device roadmap for electronic industry proposed by Rathnayake.

temperature substrates. For these reasons, alternative materials including polymers, organic molecules, and more recently nanotubes and nanowires have been gaining a lot of attention for various emerging applications. The long-range crystalline order of MOFs implies that charge transport through delocalized conduction and valence bands typical of crystalline inorganic semiconductors is possible. Emergence of delocalized bands in MOFs will require that the π orbitals in the linker groups overlap effectively with the metal d orbitals. Such overlap is absent in the majority of synthetically known MOFs where carboxylate oxygen atoms are coordinated to the metal center through σ bonds. Therefore, most MOFs are electrical insulators. This barrier needs to be overcome in next decades, perhaps by synthesizing novel MOFs using higher order conjugated linkers and increasing the functional density of the MOFs. Modifying the linker structure could lead to better charge transfer between linker and the metal cations of the framework. One possible route has suggested replacing the carboxylate terminating linkers with isocyanide groups [89]. It has been shown that, Prussian Blue, a mixed valence crystalline compound with Fe(II) and Fe(III) ions coordinated with isocyanide ligands, is electrically conducting [90]. Another approach, suggested by Allendorf et al., is to introduce conducting phases into the MOF channels [89]. Some other approaches have been taken place to enhance electronic transport properties of MOFs by introducing other conductive nanomaterials, inorganic oxides, polymers, and carbon nanotubes into MOFs framework [91–93].

Atomic level fundamental understanding that cannot be obtained readily from experimental methods, is necessary to address MOFs electronic band structure, density of state, band gap, and electron and hole mobilities. There has been increasing accuracy of predictive results using molecular dynamics (MD) force fields (FF) and DFT approximations for various MOFs' property studies [94–100]. Density Functional Theory (DFT) methods using periodic boundary conditions have been popular for predicting the electronic structure of MOFs [57, 67, 101, 102]. However, DFT-based computational calculations are underestimate excited state energies by about a factor of two. Adapting high-accuracy methods, such as Quantum Monte Carlo (QMC), DFT + U, and GW are not feasible for systems with large numbers of electrons. For example, practical QMC calculations currently could not apply to the systems that exceed 1000 electrons. One formula unit of IRMOF-1 has 760 electrons and 106 atoms. Owing to these limitations in current computational approaches, MOFs are much more challenging than traditional electronic materials with much smaller unit cells. The computational methods for predicting properties of MOFs are at an early stage of development, in particularly for predicting electronic properties of MOFs [57, 101]. Developing simple and rapid analytical approaches are not only a necessary tool to experimental investigations but also can be used as an accelerate investigation and predictive tool by themselves to screen semiconducting MOFs from the database of synthetically known MOFs. Such analytical approaches combined with computational studies will eventually enable the design of machine learning approaches for large-scale screening of existing and hypothetical MOF structures for specific applications [103–105].

MOFs are also showing promise in their use as electrolytes due to their low electronic conductivity, tunable polarity, and high porosity [106]. There are many ways that MOFs have been employed to elevate the downfalls of current electrolytes. For example, they have been using as hosts for liquid electrolyte solutions or ionic liquids [107, 108]. However, the drying of the electrolyte solution within the MOFs presents an issue since the ion transport is mostly achieved by the solvent molecules within the electrolyte rather than by the MOF itself. Furthermore, MOFs is used as a filler to reduce the crystallinity of SPEs [107, 108]. However, up to date MOFs have not been explored to be used as a solid electrolyte excepts in a composite form [109].

One way to achieve this is designing lithium-based metal organic frameworks (Li MOF) where excess lithium is transferred through the defects in the MOF structure. However, research regarding Li MOFs as solid electrolytes is currently lacking. The majority of MOF/electrolyte studies are only focused on employing MOFs as a host of ionically conductive materials rather than utilizing MOFs as solid-state electrolytes. Therefore, we identified this area of research in the proposed roadmap to stimulate investigating the potential of Li-based MOFs as solid electrolytes. There are different types of Li-MOFs already developed [110–112], but many of them are designed for applications other than battery electrolytes. We believe that Li- MOF structures can be tuned for lithium transport. Overall, Li MOFs show potential for the use as solid ionic conductors and much research should be performed to explore their possibility for solid state electrolytes and battery components.

Exploring thermoelectric properties of MOFs emerges five years ago along with exploring the electronic properties of MOFs by systematic structural modifications and introducing guest molecules onto MOFs. The first thermoelectric property measurements on MOFs has introduced by Erikson in 2015 [113]. then, up to date, there have been less than ten publications in thermoelectric MOFs, thus this field of research is relatively new. Highly nanoporous MOFs are promising since porosity can reduce the lattice thermal conductivity. The effect the conjugation length of the organic linker that tailors the pore dimension for lattice thermal conductivity must be investigated. The thermoelectric figure of merit that measures the efficiency of a thermoelectric device can be improved by decreasing the lattice thermal conductivity. It is believed that changing the conjugation length or the complexity of the organic linker changes phonon scattering, thereby changing the lattice thermal conductivity [77, 114]. The ligand modifications can be successfully achieved by isoreticular synthesis approaches. Also, increasing the porosity of MOFs increases phonon scattering that also reduces thermal conductivity [114]. Therefore, in order to utilize MOFs as active materials in thermoelectric devices, understanding the contribution of phonon vibrations to lattice thermal conductivity is essential and must be investigated. Directing future research on thermoelectric MOFs towards experimentally investigating thermoelectric properties of MOF based thin films to find ways of decreasing thermal conductivity by structural modifications to the organic ligand is beneficial.

In order to use MOFs as photoactive layer for energy harvesting and conversion, MOFs should possess decent light harvesting capability in the region from visible light to near-infrared (NIR). As the material's light-harvesting window is primarily determined by its band gap, synthesizing a MOF with a semiconducting band gap that can absorb light in the solar spectrum should be a requirement for it to serve as the photoactive material. Given that the electronic configuration of MOFs is contributed by both the constituent metal ion and the organic linker, the resultant bandgap and semiconducting properties of MOFs can thus be tailored by their structural design and engineering. Since most MOFs possess large band gap due to lack of overlap between metal ion and the organic linker and low degree of conjugation, they cannot effectively absorb light in the solar spectrum. The ligand center of MOFs plays a dominant role in its resulting light harvesting behavior [77, 114]. Tailoring the structure and its composition, MOFs charge transfer processes can be improved to enable the photocurrent of MOFs and fulfilling the photoactive functions.

To effectively reduce the band gap of MOFs and enrich their semiconducting properties for photovoltaic applications, three strategies can be implemented and have been identified [115]. These strategies are: (1) selecting electron rich metal nodes and conjugated-based organic molecules, (2) enhancing the conjugation of the organic linker, and (3) functionalizing the organic linker with electron-donating groups, such as hydroxyl, nitro, and amino groups. Additionally, facilitating electron delocalization through guest-mediated p-donor/acceptor stacks can also effectively diminish the band gaps of the materials [115]. Besides narrowing the band gap, electronic structure that contributes the semiconducting properties of MOFs also play a vital role as sufficient dissociation of the photoexcitons generated in the MOFs is required to produce a reasonable photocurrent. In this regard, MOFs exhibit critical barriers to use as the photoactive materials directly and impedes its progress in photovoltaic applications to date. However, up to date, besides acting as the photoactive materials, the MOFs has been contributing to the photovoltaic community by serving as functional additives or interlayers to improve the performance and stability of the derived solar cell devices. In order to utilize MOFs for photoactive layer in photovoltaics, it is necessary to design electrically conductive MOFs. The research efforts developing more functional conducting MOFs are required in the coming decade.

5. Conclusions

Owing to synthetic processability using reticular chemistry, MOFs offer unusual properties paving the path for many opportunities and their use in optoelectronic devices. Their use in devices so far is limited to sensors and gas storage. However, MOFs field is moving towards exploring their optical and electrical properties to use in electronic devices. There are many MOFs with tunable bandgap, both ultralow-k and high-k dielectric constants, varied magnetic properties, luminescence, and a few with semiconducting behavior, suggesting MOFs as emerging material with unique properties exceeding any other class of materials. Combining the solvothermal synthesis method with self-assembly processes, we can achieve highly ordered nanoporous structures with precise dimensionality that creates the potential for electronics and self-assembly with atomic-scale resolution and precision. In order to become MOFs for electronic devices, many challenges must be solved, and electronic structures of MOFs should be revealed. The MOFs-device roadmap should be one meaningful way to reach MOFs milestones for optoelectronic devices and will enable MOFs to be performed in their best, as well as allowing the necessary integration with other materials to fabricate fully functional devices.

Acknowledgements

Authors acknowledge the Joint School of Nanoscience and Nanoengineering, a member of the South-eastern Nanotechnology Infrastructure Corridor (SENIC) and National Nanotechnology Coordinated Infrastructure (NNCI), supported by the NSF (Grant ECCS-1542174).

Conflict of interest

There is no conflict of interest to declare.

Author details

Hemali Rathnayake* and Sheeba Dawood
Department of Nanoscience, Joint School of Nanoscience and Nanoengineering, University of North Carolina at Greensboro, Greensboro, NC, USA

*Address all correspondence to: hprathna@uncg.edu

References

[1] Zafar F, Sharmin E (2016) Metal-Organic Frameworks. BoD–Books on Demand

[2] Ogawa T, Iyoki K, Fukushima T, Kajikawa Y (2017) Landscape of research areas for zeolites and metal-organic frameworks using computational classification based on citation networks. Materials 10:1428

[3] Furukawa H, Cordova KE, O'Keeffe M, Yaghi OM (2013) The chemistry and applications of metal-organic frameworks. Science 341:

[4] Eslava S, Zhang L, Esconjauregui S, Yang J, Vanstreels K, Baklanov MR, Saiz E (2013) Metal-organic framework ZIF-8 films as low-κ dielectrics in microelectronics. Chemistry of Materials 25:27-33

[5] Yap MH, Fow KL, Chen GZ (2017) Synthesis and applications of MOF-derived porous nanostructures. Green Energy & Environment 2:218-245

[6] Logar NZ, Kaucic V (2006) Nanoporous materials: from catalysis and hydrogen storage to wastewater treatment. Acta chimica slovenica 53:117

[7] Espallargas GM, Coronado E (2018) Magnetic functionalities in MOFs: from the framework to the pore. Chemical Society Reviews 47:533-557

[8] Vikrant K, Kumar V, Kim K-H, Kukkar D (2017) Metal–organic frameworks (MOFs): potential and challenges for capture and abatement of ammonia. Journal of Materials Chemistry A 5:22877-22896

[9] Ma M, Lu L, Li H, Xiong Y, Dong F (2019) Functional metal organic framework/sio2 nanocomposites: From versatile synthesis to advanced applications. Polymers 11:1823

[10] Liu Y, Zhao Y, Chen X (2019) Bioengineering of metal-organic frameworks for nanomedicine. Theranostics 9:3122

[11] Rowsell JL, Yaghi OM (2004) Metal–organic frameworks: a new class of porous materials. Microporous and mesoporous materials 73:3-14

[12] Kotzabasaki M, Froudakis GE (2018) Review of computer simulations on anti-cancer drug delivery in MOFs. Inorganic Chemistry Frontiers 5:1255-1272

[13] Rowsell JL, Spencer EC, Eckert J, Howard JA, Yaghi OM (2005) Gas adsorption sites in a large-pore metal-organic framework. Science 309:1350-1354

[14] Yang J, Trickett CA, Alahmadi SB, Alshammari AS, Yaghi OM (2017) Calcium L-lactate frameworks as naturally degradable carriers for pesticides. Journal of the American Chemical Society 139:8118-8121

[15] Silva CG, Corma A, García H (2010) Metal–organic frameworks as semiconductors. Journal of Materials Chemistry 20:3141-3156

[16] Stassen I, Burtch N, Talin A, Falcaro P, Allendorf M, Ameloot R (2017) An updated roadmap for the integration of metal–organic frameworks with electronic devices and chemical sensors. Chemical Society Reviews 46:3185-3241

[17] Usman M, Mendiratta S, Lu K-L (2017) Semiconductor metal–organic frameworks: Future low-bandgap materials. Advanced Materials 29:1605071

[18] Stavila V, Talin AA, Allendorf MD (2014) MOF-based electronic and opto-electronic devices. Chemical Society Reviews 43:5994-6010

[19] Guo Z, Panda DK, Gordillo MA, Khatun A, Wu H, Zhou W, Saha S (2017) Lowering Band Gap of an Electroactive Metal–Organic Framework via Complementary Guest Intercalation. ACS applied materials & interfaces 9:32413-32417

[20] Deng X, Hu J-Y, Luo J, Liao W-M, He J (2020) Conductive Metal–Organic Frameworks: Mechanisms, Design Strategies and Recent Advances. Topics in Current Chemistry 378:1-50

[21] Allendorf MD, Bauer CA, Bhakta RK, Houk RJT (2009) Luminescent metal–organic frameworks. Chemical Society Reviews 38:1330-1352

[22] Yang C, Dong R, Wang M, Petkov PS, Zhang Z, Wang M, Han P, Ballabio M, Bräuninger SA, Liao Z (2019) A semiconducting layered metal-organic framework magnet. Nature communications 10:1-9

[23] Dawood S, Yarbrough R, Davis K, Rathnayake H (2019) Self-assembly and optoelectronic properties of isoreticular MOF nanocrystals. Synthetic Metals 252:107-112

[24] Rowsell JL, Yaghi OM (2004) Metal–organic frameworks: a new class of porous materials. Microporous and mesoporous materials 73:3-14

[25] Jhung SH, Lee J-H, Yoon JW, Serre C, Férey G, Chang J-S (2007) Microwave synthesis of chromium terephthalate MIL-101 and its benzene sorption ability. Advanced Materials 19:121-124

[26] Castaldelli E, Imalka Jayawardena KDG, Cox DC, Clarkson GJ, Walton RI, Le-Quang L, Chauvin J, Silva SRP, Demets GJ-F (2017) Electrical semiconduction modulated by light in a cobalt and naphthalene diimide metal-organic framework. Nat Commun 8:2139

[27] Qiu L-G, Xu T, Li Z-Q, Wang W, Wu Y, Jiang X, Tian X-Y, Zhang L-D (2008) Hierarchically micro-and mesoporous metal–organic frameworks with tunable porosity. Angewandte Chemie International Edition 47:9487-9491

[28] Peng L, Zhang J, Xue Z, Han B, Sang X, Liu C, Yang G (2014) Highly mesoporous metal–organic framework assembled in a switchable solvent. Nat Commun 5:4465

[29] Zhao Y, Zhang J, Han B, Song J, Li J, Wang Q (2011) Metal–organic framework nanospheres with well-ordered mesopores synthesized in an ionic liquid/CO2/surfactant system. Angewandte Chemie International Edition 50:636-639

[30] Kung C-W, Han P-C, Chuang C-H, Wu KC-W (2019) Electronically conductive metal–organic framework-based materials. APL Materials 7:110902

[31] Goswami S, Ray D, Otake K, Kung C-W, Garibay SJ, Islamoglu T, Atilgan A, Cui Y, Cramer CJ, Farha OK (2018) A porous, electrically conductive hexa-zirconium (IV) metal–organic framework. Chemical science 9:4477-4482

[32] Sun L, Hendon CH, Minier MA, Walsh A, Dincă M (2015) Million-fold electrical conductivity enhancement in Fe2 (DEBDC) versus Mn2 (DEBDC)(E= S, O). Journal of the American Chemical Society 137:6164-6167

[33] Talin AA, Centrone A, Ford AC, Foster ME, Stavila V, Haney P, Kinney RA, Szalai V, El Gabaly F, Yoon HP (2014) Tunable electrical conductivity in metal-organic framework thin-film devices. Science 343:66-69

[34] Banerjee S, Tyagi AK (2011) Functional materials: preparation, processing and applications. Elsevier

[35] Wu G, Huang J, Zang Y, He J, Xu G (2017) Porous field-effect transistors based on a semiconductive metal–organic framework. Journal of the American Chemical Society 139:1360-1363

[36] Narayan TC, Miyakai T, Seki S, Dincă M (2012) High charge mobility in a tetrathiafulvalene-based microporous metal–organic framework. Journal of the American Chemical Society 134:12932-12935

[37] Kobayashi Y, Jacobs B, Allendorf MD, Long JR (2010) Conductivity, Doping, and Redox Chemistry of a Microporous Dithiolene-Based Metal–Organic Framework. Chem Mater 22:4120-4122

[38] Pathak A, Shen J-W, Usman M, Wei L-F, Mendiratta S, Chang Y-S, Sainbileg B, Ngue C-M, Chen R-S, Hayashi M (2019) Integration of a (–Cu–S–) n plane in a metal–organic framework affords high electrical conductivity. Nature communications 10:1-7

[39] Li H, Eddaoudi M, O'Keeffe M, Yaghi OM (1999) Design and synthesis of an exceptionally stable and highly porous metal-organic framework. nature 402:276-279

[40] Mueller T, Ceder G (2005) A density functional theory study of hydrogen adsorption in MOF-5. The Journal of Physical Chemistry B 109:17974-17983

[41] Bordiga S, Vitillo JG, Ricchiardi G, Regli L, Cocina D, Zecchina A, Arstad B, Bjørgen M, Hafizovic J, Lillerud KP (2005) Interaction of hydrogen with MOF-5. The Journal of Physical Chemistry B 109:18237-18242

[42] Li Y, Yang RT (2006) Significantly enhanced hydrogen storage in metal–organic frameworks via spillover. Journal of the American Chemical Society 128:726-727

[43] Hafizovic J, Bjørgen M, Olsbye U, Dietzel PD, Bordiga S, Prestipino C, Lamberti C, Lillerud KP (2007) The inconsistency in adsorption properties and powder XRD data of MOF-5 is rationalized by framework interpenetration and the presence of organic and inorganic species in the nanocavities. Journal of the American Chemical Society 129:3612-3620

[44] Huang L, Wang H, Chen J, Wang Z, Sun J, Zhao D, Yan Y (2003) Synthesis, morphology control, and properties of porous metal–organic coordination polymers. Microporous and mesoporous materials 58:105-114

[45] Tranchemontagne DJ, Hunt JR, Yaghi OM (2008) Room temperature synthesis of metal-organic frameworks: MOF-5, MOF-74, MOF-177, MOF-199, and IRMOF-0. Tetrahedron 64:8553-8557

[46] Bordiga S, Lamberti C, Ricchiardi G, Regli L, Bonino F, Damin A, Lillerud K-P, Bjorgen M, Zecchina A (2004) Electronic and vibrational properties of a MOF-5 metal–organic framework: ZnO quantum dot behaviour. Chemical communications 2300-2301

[47] Tachikawa T, Choi JR, Fujitsuka M, Majima T (2008) Photoinduced Charge-Transfer Processes on MOF-5 Nanoparticles: Elucidating Differences between Metal-Organic Frameworks and Semiconductor Metal Oxides. J Phys Chem C 112:14090-14101

[48] Stock N, Biswas S (2012) Synthesis of Metal-Organic Frameworks (MOFs): Routes to Various MOF Topologies, Morphologies, and Composites. Chem Rev 112:933-969

[49] Sun L, Campbell MG, Dincă M (2016) Electrically Conductive Porous Metal-Organic Frameworks. Angew Chem Int Ed 55:3566-3579

[50] Eddaoudi M, Kim J, Rosi N, Vodak D, Wachter J, O'Keeffe M, Yaghi OM (2002) Systematic Design of Pore Size and Functionality in Isoreticular MOFs and Their Application in Methane Storage. Science 295:469-472

[51] Rowsell JLC, Millward AR, Park KS, Yaghi OM (2004) Hydrogen Sorption in Functionalized Metal–Organic Frameworks. J Am Chem Soc 126:5666-5667

[52] Yao Q, Su J, Cheung O, Liu Q, Hedin N, Zou X (2012) Interpenetrated metal–organic frameworks and their uptake of CO2 at relatively low pressures. J Mater Chem 22:10345

[53] Das MC, Xu H, Wang Z, Srinivas G, Zhou W, Yue Y-F, Nesterov VN, Qian G, Chen B (2011) A Zn4O-containing doubly interpenetrated porous metal–organic framework for photocatalytic decomposition of methyl orange. Chem Commun 47:11715

[54] Perry IV JJ, Feng PL, Meek ST, Leong K, Doty FP, Allendorf MD (2012) Connecting structure with function in metal–organic frameworks to design novel photo- and radioluminescent materials. J Mater Chem 22:10235

[55] Feldblyum JI, Wong-Foy AG, Matzger AJ (2012) Non-interpenetrated IRMOF-8: synthesis, activation, and gas sorption. Chem Commun 48:9828

[56] Dong W, Sun Y-Q, Yu B, et al (2004) Synthesis, crystal structures and luminescent properties of two supramolecular assemblies containing [Au(CN)2]– building block. New J Chem 28:1347-1351

[57] Pham HQ, Mai T, Pham-Tran N-N, Kawazoe Y, Mizuseki H, Nguyen-Manh D (2014) Engineering of Band Gap in Metal–Organic Frameworks by Functionalizing Organic Linker: A Systematic Density Functional Theory Investigation. J Phys Chem C 118:4567-4577

[58] Li H, Eddaoudi M, Groy TL, Yaghi OM (1998) Establishing Microporosity in Open Metal–Organic Frameworks: Gas Sorption Isotherms for Zn(BDC) (BDC = 1,4-Benzenedicarboxylate). J Am Chem Soc 120:8571-8572

[59] Eddaoudi M, Moler DB, Li H, Chen B, Reineke TM, O'Keeffe M, Yaghi OM (2001) Modular Chemistry: Secondary Building Units as a Basis for the Design of Highly Porous and Robust Metal–Organic Carboxylate Frameworks. Acc Chem Res 34:319-330

[60] Ockwig NW, Delgado-Friedrichs O, O'Keeffe M, Yaghi OM (2005) Reticular Chemistry: Occurrence and Taxonomy of Nets and Grammar for the Design of Frameworks. Acc Chem Res 38:176-182

[61] Xiong R, Odbadrakh K, Michalkova A, Luna JP, Petrova T, Keffer DJ, Nicholson DM, Fuentes-Cabrera MA, Lewis JP, Leszczynski J (2010) Evaluation of functionalized isoreticular metal organic frameworks (IRMOFs) as smart nanoporous preconcentrators of RDX. Sensors and Actuators B: Chemical 148:459-468

[62] Odbadrakh K, Lewis JP, Nicholson DM, Petrova T, Michalkova A, Leszczynski J (2010) Interactions of Cyclotrimethylene Trinitramine (RDX) with Metal–Organic Framework IRMOF-1. J Phys Chem C 114:3732-3736

[63] Petrova T, Michalkova A, Leszczynski J (2010) Adsorption of RDX and TATP on IRMOF-1: an ab initio study. Struct Chem 21:391-404

[64] Xiong R, Keffer DJ, Fuentes-Cabrera M, Nicholson DM, Michalkova A, Petrova T, Leszczynski J, Odbadrakh K, Doss BL, Lewis JP (2010) Effect of Charge Distribution on RDX Adsorption in IRMOF-10. Langmuir 26:5942-5950

[65] Deshpande RK, Minnaar JL, Telfer SG (2010) Thermolabile

Groups in Metal-Organic Frameworks: Suppression of Network Interpenetration, Post-Synthetic Cavity Expansion, and Protection of Reactive Functional Groups. Angew Chem Int Ed 49:4598-4602

[66] Yang L-M, Ravindran P, Vajeeston P, Tilset M (2012) Ab initio investigations on the crystal structure, formation enthalpy, electronic structure, chemical bonding, and optical properties of experimentally synthesized isoreticular metal–organic framework-10 and its analogues: M-IRMOF-10 (M = Zn, Cd, Be, Mg, Ca, Sr and Ba). RSC Adv 2:1618-1631

[67] Gascon J, Hernández-Alonso MD, Almeida AR, van Klink GPM, Kapteijn F, Mul G (2008) Isoreticular MOFs as Efficient Photocatalysts with Tunable Band Gap: An Operando FTIR Study of the Photoinduced Oxidation of Propylene. ChemSusChem 1:981-983

[68] Houk RJT, Jacobs BW, Gabaly FE, Chang NN, Talin AA, Graham DD, House SD, Robertson IM, Allendorf MD (2009) Silver Cluster Formation, Dynamics, and Chemistry in Metal–Organic Frameworks. Nano Lett 9:3413-3418

[69] Cao A, Zhu W, Shang J, Klootwijk JH, Sudhölter EJR, Huskens J, de Smet LCPM (2017) Metal–Organic Polyhedra-Coated Si Nanowires for the Sensitive Detection of Trace Explosives. Nano Lett 17:1-7

[70] Aubrey ML, Ameloot R, Wiers BM, Long JR (2014) Metal–organic frameworks as solid magnesium electrolytes. Energy Environ Sci 7:667-671

[71] Sheberla D, Sun L, Blood-Forsythe MA, Er S, Wade CR, Brozek CK, Aspuru-Guzik A, Dincă M (2014) High Electrical Conductivity in Ni3(2,3,6,7,10,11-hexaiminotriphenylene)2, a Semiconducting Metal–Organic Graphene Analogue. J Am Chem Soc 136:8859-8862

[72] Schröder M (ed) (2010) Functional metal-organic frameworks: gas storage, separation, and catalysis. Springer, Berlin ; New York

[73] Li J-R, Kuppler RJ, Zhou H-C (2009) Selective gas adsorption and separation in metal–organic frameworks. Chem Soc Rev 38:1477-1504

[74] Lee J, Farha OK, Roberts J, Scheidt KA, Nguyen ST, Hupp JT (2009) Metal–organic framework materials as catalysts. Chem Soc Rev 38:1450-1459

[75] Alvaro M, Carbonell E, Ferrer B, Llabrés i Xamena FX, Garcia H (2007) Semiconductor Behavior of a Metal-Organic Framework (MOF). Chem Eur J 13:5106-5112

[76] Takaishi S, Hosoda M, Kajiwara T, Miyasaka H, Yamashita M, Nakanishi Y, Kitagawa Y, Yamaguchi K, Kobayashi A, Kitagawa H (2009) Electroconductive Porous Coordination Polymer Cu[Cu(pdt)2] Composed of Donor and Acceptor Building Units. Inorg Chem 48:9048-9050

[77] Civalleri B, Napoli F, Noël Y, Roetti C, Dovesi R (2006) Ab-initio prediction of materials properties with CRYSTAL: MOF-5 as a case study. CrystEngComm 8:364-371

[78] Fuentes-Cabrera M, Nicholson DM, Sumpter BG, Widom M (2005) Electronic structure and properties of isoreticular metal-organic frameworks: The case of M-IRMOF1 (M=Zn, Cd, Be, Mg, and Ca). The Journal of Chemical Physics 123:124713

[79] Kuc A, Enyashin A, Seifert G (2007) Metal–Organic Frameworks: Structural, Energetic, Electronic, and Mechanical Properties. J Phys Chem B 111:8179-8186

[80] Kurmoo M (2009) Magnetic metal–organic frameworks. Chem Soc Rev 38:1353-1379

[81] Guo Z, Cao R, Wang X, Li H, Yuan W, Wang G, Wu H, Li J (2009) A Multifunctional 3D Ferroelectric and NLO-Active Porous Metal–Organic Framework. J Am Chem Soc 131:6894-6895

[82] Xie Y-M, Liu J-H, Wu X-Y, Zhao Z-G, Zhang Q-S, Wang F, Chen S-C, Lu C-Z (2008) New Ferroelectric and Nonlinear Optical Porous Coordination Polymer Constructed from a Rare (CuBr)∞ Castellated Chain. Crystal Growth & Design 8:3914-3916

[83] Bureekaew S, Horike S, Higuchi M, Mizuno M, Kawamura T, Tanaka D, Yanai N, Kitagawa S (2009) One-dimensional imidazole aggregate in aluminium porous coordination polymers with high proton conductivity. Nature Mater 8:831-836

[84] Hurd JA, Vaidhyanathan R, Thangadurai V, Ratcliffe CI, Moudrakovski IL, Shimizu GKH (2009) Anhydrous proton conduction at 150 °C in a crystalline metal–organic framework. Nature Chem 1:705-710

[85] Sadakiyo M, Yamada T, Kitagawa H (2009) Rational Designs for Highly Proton-Conductive Metal–Organic Frameworks. J Am Chem Soc 131:9906-9907

[86] Taylor JM, Mah RK, Moudrakovski IL, Ratcliffe CI, Vaidhyanathan R, Shimizu GKH (2010) Facile Proton Conduction via Ordered Water Molecules in a Phosphonate Metal–Organic Framework. J Am Chem Soc 132:14055-14057

[87] Fletcher AJ, Thomas KM, Rosseinsky MJ (2005) Flexibility in metal-organic framework materials: Impact on sorption properties. Journal of Solid State Chemistry 178:2491-2510

[88] Uemura K, Matsuda R, Kitagawa S (2005) Flexible microporous coordination polymers. Journal of Solid State Chemistry 178:2420-2429

[89] Allendorf MD, Schwartzberg A, Stavila V, Talin AA (2011) A Roadmap to Implementing Metal-Organic Frameworks in Electronic Devices: Challenges and Critical Directions. Chem Eur J 17:11372-11388

[90] Behera JN, D'Alessandro DM, Soheilnia N, Long JR (2009) Synthesis and Characterization of Ruthenium and Iron–Ruthenium Prussian Blue Analogues. Chem Mater 21:1922-1926

[91] Sosa J, Bennett T, Nelms K, Liu B, Tovar R, Liu Y (2018) Metal–Organic Framework Hybrid Materials and Their Applications. Crystals 8:325

[92] Zhang Z, Nguyen HTH, Miller SA, Cohen SM (2015) polyMOFs: A Class of Interconvertible Polymer-Metal-Organic-Framework Hybrid Materials. Angew Chem Int Ed 54:6152-6157

[93] Parnham ER, Morris RE (2007) Ionothermal Synthesis of Zeolites, Metal–Organic Frameworks, and Inorganic–Organic Hybrids. Acc Chem Res 40:1005-1013

[94] Tafipolsky M, Amirjalayer S, Schmid R (2007) Ab initio parametrized MM3 force field for the metal-organic framework MOF-5. J Comput Chem 28:1169-1176

[95] Bureekaew S, Amirjalayer S, Tafipolsky M, Spickermann C, Roy TK, Schmid R (2013) MOF-FF - A flexible first-principles derived force field for metal-organic frameworks: MOF-FF - A force field for metal-organic frameworks. Phys Status Solidi B 250:1128-1141

[96] Han SS, Choi S-H, van Duin ACT (2010) Molecular dynamics simulations of stability of metal–organic frameworks against H2O using the

ReaxFF reactive force field. Chem Commun 46:5713-5715

[97] Yang Q, Liu D, Zhong C, Li J-R (2013) Development of Computational Methodologies for Metal–Organic Frameworks and Their Application in Gas Separations. Chem Rev 113:8261-8323

[98] Odoh SO, Cramer CJ, Truhlar DG, Gagliardi L (2015) Quantum-Chemical Characterization of the Properties and Reactivities of Metal–Organic Frameworks. Chem Rev 115:6051-6111

[99] Grajciar L, Wiersum AD, Llewellyn PL, Chang J-S, Nachtigall P (2011) Understanding CO2 Adsorption in CuBTC MOF: Comparing Combined DFT–ab Initio Calculations with Microcalorimetry Experiments. J Phys Chem C 115:17925-17933

[100] Chavan S, Vitillo JG, Gianolio D, et al (2012) H2storage in isostructural UiO-67 and UiO-66 MOFs. Phys Chem Chem Phys 14:1614-1626

[101] Kumar A, Banerjee K, Foster AS, Liljeroth P (2018) Two-Dimensional Band Structure in Honeycomb Metal–Organic Frameworks. Nano Lett 18:5596-5602

[102] Li J, Musho T, Bright J, Wu N (2019) Functionalization of a Metal-Organic Framework Semiconductor for Tuned Band Structure and Catalytic Activity. J Electrochem Soc 166:H3029–H3034

[103] Weitzner SE, Dabo I (2017) Quantum–continuum simulation of underpotential deposition at electrified metal–solution interfaces. npj Comput Mater 3:1

[104] Wilmer CE, Leaf M, Lee CY, Farha OK, Hauser BG, Hupp JT, Snurr RQ (2012) Large-scale screening of hypothetical metal–organic frameworks. Nature Chem 4:83-89

[105] Witman M, Ling S, Anderson S, Tong L, Stylianou KC, Slater B, Smit B, Haranczyk M (2016) In silico design and screening of hypothetical MOF-74 analogs and their experimental synthesis. Chem Sci 7:6263-6272

[106] Zhao R, Liang Z, Zou R, Xu Q (2018) Metal-Organic Frameworks for Batteries. Joule 2:2235-2259

[107] Fu X, Yu D, Zhou J, Li S, Gao X, Han Y, Qi P, Feng X, Wang B (2016) Inorganic and organic hybrid solid electrolytes for lithium-ion batteries. CrystEngComm 18:4236-4258

[108] Wiers BM, Foo M-L, Balsara NP, Long JR (2011) A Solid Lithium Electrolyte via Addition of Lithium Isopropoxide to a Metal–Organic Framework with Open Metal Sites. J Am Chem Soc 133:14522-14525

[109] Yuan C, Li J, Han P, Lai Y, Zhang Z, Liu J (2013) Enhanced electrochemical performance of poly(ethylene oxide) based composite polymer electrolyte by incorporation of nano-sized metal-organic framework. Journal of Power Sources 240:653-658

[110] Zhao X, Wu T, Zheng S-T, Wang L, Bu X, Feng P (2011) A zeolitic porous lithium–organic framework constructed from cubane clusters. Chem Commun 47:5536-5538

[111] Banerjee D, Borkowski LA, Kim SJ, Parise JB (2009) Synthesis and Structural Characterization of Lithium-Based Metal−Organic Frameworks. Crystal Growth & Design 9:4922-4926

[112] Ogihara N, Ohba N, Kishida Y (2017) On/off switchable electronic conduction in intercalated metal-organic frameworks. Sci Adv 3:e1603103

[113] Erickson KJ, Léonard F, Stavila V, Foster ME, Spataru CD, Jones RE, Foley BM, Hopkins PE, Allendorf MD, Talin AA (2015) Thin

Film Thermoelectric Metal-Organic Framework with High Seebeck Coefficient and Low Thermal Conductivity. Adv Mater 27:3453-3459

[114] Yang L-M, Vajeeston P, Ravindran P, Fjellvåg H, Tilset M (2010) Theoretical Investigations on the Chemical Bonding, Electronic Structure, And Optical Properties of the Metal–Organic Framework MOF-5. Inorg Chem 49:10283-10290

[115] Chueh C-C, Chen C-I, Su Y-A, Konnerth H, Gu Y-J, Kung C-W, Wu KC-W (2019) Harnessing MOF materials in photovoltaic devices: recent advances, challenges, and perspectives. J Mater Chem A 7:17079-17095